FLY TYING
FOR EVERYONE

FLY TYING
FOR EVERYONE

TIM CAMMISA

STACKPOLE
BOOKS
Guilford, Connecticut

Published by Stackpole Books
An imprint of The Rowman & Littlefield Publishing Group, Inc.
4501 Forbes Blvd., Ste. 200
Lanham, MD 20706
www.rowman.com

Distributed by NATIONAL BOOK NETWORK

British Library Cataloguing in Publication Information available

Library of Congress Cataloging-in-Publication Data

Names: Cammisa, Tim, 1979– author.
Title: Fly tying for everyone / Tim Cammisa.
Description: Guilford, Connecticut : Stackpole Books, [2021] | Summary:
  "Author Tim Cammisa teaches readers how to tie these simple but
  effective patterns and then how to take the techniques they've learned
  and use them for most of the other core patterns—old and new—that should
  be in every angler's box. Includes information on the latest materials,
  tying tips from other tiers, and 13 patterns with recipes and complete
  step-by-step instructions"— Provided by publisher.
Identifiers: LCCN 2020058424 (print) | LCCN 2020058425 (ebook) | ISBN
  9780811738859 (hardback) | ISBN 9780811768900 (epub)
Subjects: LCSH: Fly tying.
Classification: LCC SH451 .C29 2021 (print) | LCC SH451 (ebook) | DDC
  688.7/9124—dc23
LC record available at https://lccn.loc.gov/2020058424
LC ebook record available at https://lccn.loc.gov/2020058425

ISBN 978-0-8117-3885-9 (cloth: alk. paper)
ISBN 978-0-8117-6890-0 (electronic)

♾™ The paper used in this publication meets the minimum requirements of
American National Standard for Information Sciences—Permanence of Paper for
Printed Library Materials, ANSI/NISO Z39.48-1992.

For Heather, Angelo, and Josephine.
Here's to more days together chasing fish.

CONTENTS

ACKNOWLEDGMENTS

Writing this book was by no means an independent task. Support from my family and friends made everything go smoother, but I also need to acknowledge those who have pushed me to get to this point in my life. As I neared the completion of the book, it became obvious that the contributing groups were more than just those in the fly-fishing and tying community. From teachers who encouraged to mentors that led, this book is due to all of you. Thank you for the guidance and time to make me a lifelong learner and give back to others.

Most importantly, thank you to the support of my amazing and incredible wife, Heather. Comically referring to herself as a "YouTube Wife," Heather has given me love, guidance, vision, and motivation through so many moments of our life, especially the days when the macro photos just weren't coming out correctly. Couple this with the craziness of our four-year-old son Angelo (aka Energizer Bunny) and daughter Fina, born during this book's writing, and you can imagine why she's the love of my life. An outdoor enthusiast herself, Heather decided to take up fly fishing; as expected, she is a natural and always seems to catch the "big fish" every time we're on the water. From our second date to that special waterfall up to our weekend getaways to Central Pennsylvania to chase wild brown trout, Heather, I love you and thank you for being my partner in this journey, and the most important person in my life. I *promise* that I won't start writing another book for at least six months!

Suzie Cammisa, my mom, is such a positive force for our family, and her belief in me succeeding exceeded my own at times, and it's for that I'll always be grateful. Joe Cammisa, my dad, is the outdoor enthusiast that pushed me in many areas of life, and it's no wonder he is an accomplished field trialer with pointing dogs. From signing me up for that fly-tying class when I was 10, to providing all the gear, trips, and support that an impressionable kiddo could need, you've both been exceptional role models for me. You've instilled in me the value of family and support, which I plan on instilling in my children. I still don't know how you put up with the smell of mothballs in your house for all those years!

Jolene Fielder, my sister, may not have caught the fly-fishing bug but has certainly been along for the ride. One of my best days fly fishing was interrupted when I had to pick her up after a school event, and I believe that I was only an hour late. The trout wouldn't stop taking lightly twitched streamers! Her husband, Chris, and two boys, William and Finnegan, all enjoy fishing, and it's only a matter of time before I have my nephews tying flies and editing videos for me. I love you, J, and thanks for encouraging me to pursue my passions.

To my father-in-law Bill McQuillan, who has been my pseudo public relations manager for years, passing out business cards for me wherever he goes! He is always up for an adventure, no questions asked. Thank you for doing way more than just getting my name out there, and let's go fishing soon.

Jack Burke, my grandfather, instilled a drive in me to never give up, and one of my most cherished life memories was a day he took me fly fishing, not to learn himself, but to simply enjoy watching his grandson fish. You're my inspiration, and I miss you, Pop.

Angelo Cammisa, my grandfather, was a fisherman, taking trips when I was a little boy to faraway and exotic locations (e.g., New York for crappies and Canada for walleye). During my school days, I hopped on a different bus that took me to his house, just so he would take my friend and me fly fishing at a local reservoir. When we showed up at

Heather has definitely got the hang of fly fishing, from the wild brown trout of Central Pennsylvania, all the way to this gorgeous snook caught in Florida.

his door, he grabbed his coat and took us fishing, after a quick call to my parents explaining why I wouldn't be home after school. The Black Ghost killed it that day.

This book wouldn't have been possible had it not been for the vision of Jay Nichols. This project was his idea, and I am so fortunate to have had his guidance during its completion. More importantly, Jay's belief in me was great, and his thoughts and ideas along the way kept the project on track. Jay, thank you for all you've done, not just for me, but for fly fishers and tiers out there who have been inspired by your works.

To the local fly shops, you're the best. When it came to crunch time with this book, my local shop, International Angler, became a frequent stop when I needed that last bit of material (five times). Everyone needs to have a nearby shop on a local waterway, and the Neshannock Creek Fly Shop is mine. Then we get to the online world of fly shops, and Competitive Angler is second to none. Jake

Adamerovich and his family run that shop, and I consider him a close friend that I get to fish with every time I'm by the South Holston River.

To my Great Uncle John, John Dunn, and the rest of the Lunch Bunch (aka Liar's Club). This group included Doc Boal, Charlie Heathcote, Jerry Rhodaberger, and Big Fish Ed. Every Monday night, the group gathered to talk fly fishing and tying, preferably at a place offering free refills on the coffee. During my teenage years, I rarely missed one of these weekly gatherings, and learned about more than just fishing. Your guidance during my early fly-fishing and tying years led me down this path, and the constant pushing made me a better angler and tier. Every young fly fisher needs mentors like you.

Specifically, Uncle John and John Dunn have both been two of my best friends since I can remember. Each is strong-willed and passionate about fly fishing and tying, and to say I look up to each is an understatement. These two are my

My great-uncle John and his fly-tying room. Every square inch is packed full of tying materials and fly-fishing equipment, defining the true "man cave."

idols who inspired me to constantly examine and improve, and we have had some fun experiences along the way. Uncle John, from our days spent chasing local trout all over western Pennsylvania, to building bamboo fly rods in your basement, I've cherished them all. John Dunn, that first trip to Montana made such an impressionable moment on my life, as did that steelhead trip in subzero temperatures! The time we have spent together, and continue to spend, is valued, and I thank you for pushing me to new levels of this addiction.

Speaking of mentors, my most recent has been Chuck Furimsky, creator of the Fly Fishing Show. When we first connected in 2015 with an invitation to tie at the International Fly Tying Symposium, I had no idea how strong our friendship would become. Liking to talk nearly as much as me, we both pony for position during

conversations, with Chuck typically winning. Most know him as a fly-fishing personality; I am more familiar with his fly-tying innovation and design abilities, plus his nonstop desire to *always* be fishing. By the way, yes, he is a personality, too! While I was writing this book, Chuck has been fly fishing in the jungle, in Florida multiple times, and in Martha's Vineyard, Ocean City, Pennsylvania, Canada, Montauk, and Montana. All while turning seventy—thus you can see why I hold him in such high esteem and am honored to be a friend. Chuck, your shows have helped to push fly fishing and tying to new areas, and I am honored to be a part of them, even if you are to partially blame for my current endeavors. Now if I could just find a way to outfish you at that one lake!

To Rob Giannino for pushing me to take more "fly fishing journeys" and expand my thinking as an angler and in the world of media. Yes, you're my "brother from another mother," the big brother I never had. Here's to our friendship and *lots* more fly fishing and fun in the future, and here's to you potentially catching a larger fish than me someday (doubtful!).

To Josh Miller for encouraging me to jump down the rabbit hole of tightline nymphing. From that first time we fished together, not only was your talent obvious, but your ideas caused me to rethink many of my original ideas and beliefs about fly fishing. I'm honored to call you a friend, and here's to more road trips for shows and fishing.

As I became a fly-fishing presenter at events, one of my first memories was walking the aisle before a show and hearing the voice of Tim Flagler. Not "YouTube Tim Flagler," but Tim . . . right there! I immediately introduced myself, as his videos and instruction are the best in the world. Both Tim and his wife Joan are fixtures at the Fly Fishing Show, and I value my time with them both. Tim is just as gracious on the water as behind the vise, and our mutual love for IPAs guarantees a few more stories after fishing.

Speaking of the Fly Fishing Show, since taking over the reins, Ben Furimsky has brought these events to new heights. Ben's hard work and determination to put the shows together is like no other, and few realize everything that goes on behind the scenes to make things run smoothly. These shows have introduced me to many friends I look up to, including Theo Bakelaar, Tom Baltz, Greg Heffner, Phil Rowley, Mike Albano, Mike Hogue, Ted Patlen, Bruce Corwin, Aaron Jasper, and many others. Ben, thank you for allowing me to be part of The Fly Fishing Show (and family), and I can't wait to see where you take it in the future.

Joe Messinger is a close friend, living in nearby West Virginia. His tying skills are exceptional, as he continues the work started by his father at the bench. I had the honor of filming Joe tie some of his famous patterns, and I hope to get more captured for the world to learn from. Fishing with Joe is a treat, as he is quick to share pieces of advice, insight, tips, and more with everyone around. Joe, thank you for the friendship, and I look forward

to floating that special river of yours in West Virginia again.

To my YouTube subscribers and friends, thank you for the encouragement and support. Putting out videos for the world to view and comment on is both fun and time-consuming, but knowing my base is there motivates me to make each video better than the previous one. Being able to interact with you is part of the fun, and meeting many of you at events has been that much more special. Ryan Keene is one that turned into a friend, and his artwork is unlike any other in fly fishing. Ryan, my logo design is exceptional, thank you; now clear some time so we can go fishing.

To my *early* YouTube subscribers (i.e., family members), thanks for the feedback that was needed and necessary. Mike and Helen, you offered so many funny thoughts (yes, I probably *did* need a manicure) and, more importantly, advice that I value to this day. "Trout and Feather" didn't just materialize from thin air; Helen gets all the credit for naming rights.

There are many tiers out there on YouTube that I respect greatly, as the time and energy dedicated to sharing with others is much greater than most know. Gunnar Brammer is one of those, an exceptional tier and fisherman; here's to more conversations on the phone that always seem to stretch from 5 minutes to 45.

Devin Olsen is also found on YouTube, but most know him as one of the world's best anglers on Fly Fishing Team USA. Our connection started through the Fly Fishing Show, and though phone calls and texts always circle back to fly fishing, I think of one simple word to define Devin: competitor. Whether on his bike or in the water, Devin has a drive to constantly improve, and it's infectious. Thank you for the friendship, Devin, and I look forward to outfishing you on the water.

Another competitor from my home state is George Daniel. George's contributions to fly fishing are a major factor in its current popularity throughout the United States, and he remains humble through it all. All of George's books offer instruction, insight, and tips from start to finish, and you'll still find me attending every one of his presentations at a fly-fishing show. George, thanks for everything you've done to enhance the sport we love.

We're fortunate in fly tying and fly fishing, with so many great companies and people. I'm honored to call the following people friends: Alessio and Lorenzo Stonfo of Stonfo, Kevin Compton of Performance Flies, Andy and Ann Kitchener of Semperfli, Franta Hanak of Hanak, and Matthew Lourdeau of the *Casting Across* blog.

To all of my fishing and tying partners, thanks for spending time together on the water and at the bench, two of my favorite places in this world. I'm especially talking about Don Ward, and here's to more days on the Delaware fighting for time to cast dry flies to rising brown trout.

To the Seneca Valley School District, my colleagues, and students, thank you. My success on YouTube didn't happen accidentally; I'm an elementary schoolteacher used to giving daily performances . . . and feedback from students can be *much* rougher than on YouTube! My school district has additionally supported my fly-fishing endeavors, encouraging me to start a Fly Fishing and Tying Club for students. During the years, I've worked with many incredible team teachers and colleagues, with the most important, Tracey Clarke, being an inspiration for positivity.

Since I've started this project, people continually ask, "What took you so long to write a fly-tying book?" Now that I've completed one, I know why it took so long: It's more work than you can imagine! From selecting patterns that allow me to demonstrate important techniques, to knowing that each macro photograph may be scrutinized, I had to step up my game to ensure that the book would be a valuable learning tool and resource for you. For those that have come before me, I respect you more than you'll ever know. For those carrying on the torch of fly tying, I can't wait to see where you'll take it.

Thank you all for taking the time to support me and this book. Now on to the tying!

FOREWORD

It used to be for a tier to gain a reputation that there were two must-do recommendations. You could get a great start at becoming known by having an article or two appear in one of several magazines most of us subscribed to for years. *Fly Tyer* magazine was like the *Tonight Show* if you were a comedian. I couldn't wait for my issue and still get excited to see the latest patterns and materials, which amaze me because I think I've seen it all.

When I took my first fly-fishing and tying class in the early 1960s as a student at Penn State, I was lucky to register before the class was closed. I remember it as if it just happened, when I stood in a line behind a table where a staffer took student's names until the class was filled. It was in the old building called the Rec Hall, where the registration was organized so you could select the classes required to graduate in your major. I had just registered quickly in a three-person line for "Victorian Novels," which was a required course I needed for my English major. Then I hurried over to the long line to see if I could get an elective two-credit

BARRY AND CATHY BECK

xiii

class called "Techniques of Fly Fishing," taught by George Harvey. I was lucky because there were two desks left as I reached the registrar's table. No, there were no computers to complete my acceptance, you had to get in line. You couldn't call because no one had ever heard of a cell phone, and the university offices were empty because everyone was working the floor in Rec Hall. Have times changed? You bet.

As I stated in my first sentence, there were two must-do recommendations years ago to be recognized as a fly-tying celebrity. Success 101 was getting an article published in a fly-fishing or fly-tying magazine. But the second giant step was to actually write a book. Tim Cammisa is the author of *Fly Tying for Everyone*. This is his first book, but he just didn't wake up earlier in the week and decide to write a book.

Tim is part of a new generation of celebrity fly tiers. With over 30 years of fly tying, his time practicing his craft has earned him respect. When he isn't in front of his vise, he's in front of his camera. His "Trout and Feather" website has exploded, with Tim starring in over 250 videos with 23,000 subscribers and 3 million video views. You might be one of the 23,000 subscribers and like me wondered what took Tim so long to write his first book. Maybe being a schoolteacher, husband to Heather, a wife who doesn't hesitate to outfish him, and a father who backpacks his young son, Angelo, on 90 percent of his fishing trips and is now wondering how to carry his new daughter, too.

I'm not as busy as Tim, and he has gained my respect for how hard he is able to juggle so much and never drop a ball. Since I semiretired from running the Fly Fishing Show for 30 years, I have been lucky to know so many famous fly fishers and fly-tying celebrities over the years. It was fortunate that Tim and I met before my son, Ben, took over the reins of the show. I didn't have to influence Ben about continuing to feature Tim as one of the celebrity speakers. His seminars were always standing room only, and when a customer would stop me in the aisle to tell me that Tim's talk was the best one of the day, I asked Tim how much he paid his uncle to tell me that!

If you haven't met Tim yet, you'll want to once you read his book. Fishing and fly tying is definitely something everyone should try. You can learn something new every day, and if you see Tim on the stream or at a show, he'll answer any of your questions. But if he tells you about how he outfished me because my brown trout turned into a big sucker in the net, just ignore him.

—Chuck Furimsky

INTRODUCTION

Welcome to *Fly Tying for Everyone*, designed to take your tying skills to the next level and beyond. In this book you'll find patterns selected specifically for their ability to offer techniques, skills, and materials that can be easily transferred to other flies and styles. View each fly as a mini lesson, with tips and suggestions built in, both for tying and fly fishing. This book is also intended to build on your prior connections and experiences in both tying and fly fishing. Think about your own confidence patterns and why they work on the waters you fish, then enhance them with ideas from this text.

To give you an idea of what to expect, each fly is discussed in detail, accompanied by macro photography and tying steps. More importantly, I elaborate on the patterns, teasing out the critical information that drew me to select them initially. It's from this information that I challenge you to think outside the box and find ways to apply these

Catching the fish of a lifetime doesn't happen by chance, and this large wild brown trout was successfully landed on an articulated streamer pattern, using the tying and fishing styles discussed in further detail within this book. As with learning anything new, there is a learning curve; the goal of this book is to make the process much easier, leading to *your* fish of a lifetime! COURTESY OF BLACK MOUNTAIN CINEMA

techniques to your everyday patterns. This may be a simple addition to improve their durability, or an overhaul of a favorite streamer into one that's articulated; today the only limit is, honestly, your imagination.

Knowing readers crave insight and additional information, each fly description also features four extra nuggets, all intended to take your tying and fly fishing into different directions. The areas include:

1. **Tying tips.** Lots can be learned from the pictures and accompanying text, and for each fly I choose an area to focus on, intended to have an immediate impact in your tying.

2. **Featured techniques.** Highlighting a technique for each fly is something that speaks to the essence of this book. Focus on these recommendations, as mastering the basics solidifies skills that are applicable to most of today's patterns.

3. **Materials to consider.** With so many materials available on today's market, selecting them can be overwhelming. In many cases you may already have an appropriate substitute. This section will help you make additional choices for various types of patterns.

4. **Fishing suggestions.** The point of tying is to create patterns that fish want to eat, and part of the process involves understanding how flies interact with water. More importantly, *we* are an important part of that equation, and for each section, I offer suggestions for you to use when on the water.

Selecting the flies to illustrate this information was no simple task, and I drew from my own experiences in fly tying and fly fishing. Having a "traditional" start, my early tying instruction included Catskill-style dry flies from my Uncle John while fly fishing began on freestone streams and spring creeks throughout western and central Pennsylvania. As time passed and new mentors entered my life, more concepts were introduced to me, with fly-fishing experiences soon encompassing New York, California, Montana, and even Iceland!

Each mentor and location opened my awareness to a new style or water type, which in turn led to another. In my mind, fly fishing is like a scavenger hunt with no conclusion; it's up to each of us to decide how far to take it. Whether tying emergers intended to ride in the film or selecting that perfect Craft Fur color to match local baitfish, my experiences pushed me at each turn. With current fascinations including Euro nymphing and enhanced streamer techniques, there is simply no end to the learning within fly tying and fly fishing. With this book, I truly want to help you get to your next location in that hunt.

Many anglers regularly tie five or six confidence flies that produce for them on a regular basis, but unlike pattern books, this book is written to specifically push your tying skills to the next level. Whether you've just favorited a fly on Instagram, seen it hanging out of a fish's mouth in *Fly Fisherman* magazine, or read about one in the "Fly Tying with Uncle Cheech" group on Facebook, *Fly Tying for Everyone* will give you the foundation to successfully reproduce and immediately fly-fish with that pattern. Instead of shrugging a difficult fly off, use this book as a resource to understand the nuances of those techniques, which will lead both your tying and fly fishing down a new path . . . hopefully to more fish!

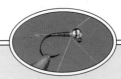

Welcome to the 21st Century of Fly Tying

The year 1989 was my first for fly tying, and I still remember the amount of materials, hook types, and patterns being honestly quite overwhelming. Having different hook models to choose from, dubbing types (in what seemed like a million colors, and every tier obviously needs them all!), plus so many feathers . . . fly tying made my head spin, yet I wanted to learn about everything. It became quickly apparent that the information had to be taken in stride, slowly integrating materials into my tying and gaining a thorough knowledge and appreciation of them one piece at a time.

Fast-forward 30 years, and it's no wonder many interested tiers skim an online fly-tying catalog and quickly make a decision to purchase a few dozen flies versus learning to tie. Material choices have multiplied immensely since my inclusion in this sport, and I feel like a student still, learning about their characteristics and the appropriate techniques to maximize each. As with most things,

Fly tying's purpose is simple: Combine materials onto a hook to persuade a fish to try it. That's it, nothing more!

great rewards include a steep learning curve, and my goal is to lessen that for all of you.

Fly tying's purpose is simple: Combine materials onto a hook to persuade a fish to try it. That's it, nothing more! The best part of this is that trout aren't even that smart, making our job even easier. Does that mean we have overcomplicated some things along the way? Possibly, but fly tying is like taking a road trip across the country—there are a nearly infinite number of routes that can get you to your destination. One way may be more efficient and faster, whereas another offers incredible scenery while taking a bit longer. This is fly tying, and the route is up to you. Knowing that may not exactly lessen the learning curve, but I'm here to help. Let's break the process down a bit and start with today's common tools.

Tier I: Essential

VISE

Vise questions are *easily* my most received piece of correspondence, and I love to say that if you ask 10 tiers for a vise recommendation, you're likely to learn about more than 15 vises! A vise is something very personal to each of us, and knowing that, I tread gently when helping others decide on the right one for their tying. Each vise offers varying options, price points, jaw sizes, and features that may be suited more for one tier versus another, and it's up to us to determine which meets our needs (and sometimes wants!).

We live in a great time to purchase a vise, with many excellent options starting around the $100 mark, then gaining traction from there. When asked about a vise, my common reply is to think about the styles and sizes of patterns commonly tied, decide on desired options for the vise, and set a budget. Finally comes the tough part, researching the various vises around that price, focusing on those that meet your daily tying needs.

When examining a vise, I first concentrate on the jaws, as I prefer those that allow me to see most of the hook, especially around the bend. This relates to my preference for tying trout flies, and that may not be critical for a tier concentrating on

articulated streamers. See how this can get tricky? Other areas that I inspect are the sections of a vise that can be adjusted, such as the jaw angles and height. Having the ability to fine-tune these areas is something I find important, as tiers have varying preferences based on hand placement, seat height, and other factors that can impact tying. Looking back, my first vise was truly fixed in all settings and didn't allow too much inspection around the hook, so I'm thankful for modern technology!

Here are some tips to help with the decision for a new vise, especially applicable if you're ready for an upgrade:

- The ability to rotate the jaws while the hook is secure is an absolute must for me. Examining the fly while tying is essential to ensure proper spacing and proportions, something I recommend every vise can perform.
- My early vises were c-clamp models, as my tying station was in a fixed location and the clamp kept the vise secure. Over the last 15 years, my needs have changed, and I bring vises around the country when fly fishing and

The first vise I tied on. Though it offered few features, it helped to fuel the addiction.

The vise featured in all tying sequences in this book is the Stonfo Transformer. An advantage of modern vises is their ability to offer many ways to customize, with this one even allowing the heads to be switched out. This image shows the traditional head, and the Transformer also comes with a streamer head and another used specifically for tube flies.

for tying events. Because of this, I now prefer a pedestal, which offers great flexibility since it's relatively easy to find a flat surface to place it on. Select your base style depending on how often you anticipate moving tying locations, and know that you can typically purchase the other attachment as an accessory.

- If interested in speed, especially for flies tied "in the round," I recommend examining vises that offer true rotary capabilities. This is a feature I choose especially when focused on tying dozens of a specific style or pattern. Tying with a rotary-style vise allows you to have complete control of the material with one hand, while the other rotates the vise at your preferred speed. Instead of wrapping materials around the hook, they remain stationary as the hook turns, which gives you the opportunity to verify their placement and spacing. Recommended vises in this category include the Stonfo Transformer, Renzetti Traveler, and Norvise.

- For ease of use, vises with spring tension are excellent choices. Few adjustments are necessary for this style, making them great choices for a range of tiers. Simply grip the arm to open the jaws and place the hook within. Consider testing the grip prior to purchase, as these can be difficult for some tiers to open. The hooks in these vises tend to be under great pressure; follow the manufacturer's instructions for proper hook placement. Recommended vises in this category include the Stonfo Kaiman and Regal series.

When searching for a vise, you may notice that the more expensive models have similar jaws as their lower-priced counterparts, even within brands. So that may leave you wondering: Why pay more? The increased cost for the expensive models may be for customization, including color choices, finishes on the vise, varying bases, and even accompanying tools. Doing your homework is no easy task with vises, and spending time on

The Renzetti Traveler was my first true rotary vise and gave me an idea of what a quality vise should offer. It is still part of my collection, and I frequently loan this vise to others who are interested in learning how to tie flies.

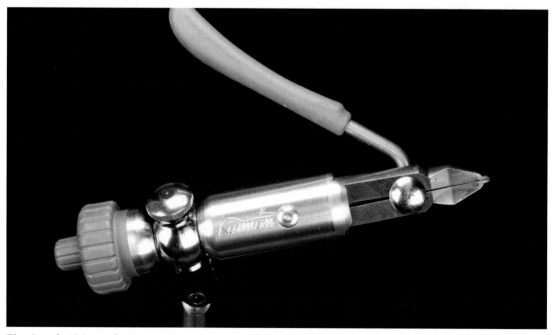

The Stonfo Kaiman features an arm to open the jaws and rotates 360 degrees. When I need to tie multiple sizes without having to worry about jaw adjustment, this is the vise I choose.

manufacturer websites will give you an idea of the differences between models.

Finally, let's talk about recommendations from others, which can further compound the selection difficulty. Being that a vise is very personal, most tiers will feel strongly about theirs, so I recommend having set questions when asking for advice. Basically, you're trying to refine their answers to help with your decision. Sample questions include:

- What styles and sizes of flies do you typically tie? Verify that their answers are similar to yours, and if so, keep asking questions!
- What other vises have you used in the past? If not many others, know that the tier may be more attached to their vise and show favoritism.
- Why would you recommend this vise to others?
- If you could change something about the vise, what would it be?
- How easy is it to adjust the vise for different hook sizes?

A final important piece to note is that there are few individuals out there who can spend time trying every vise on the market, especially those who offer recommendations via social platforms. Many fly shops will be able to offer general guidelines and can provide additional support through the process, especially as new vises on the market arrive to fill niches. Gather your thoughts and opinions from a variety of sources, then make the best decision you can with that information. When in doubt, know that the majority of today's vises are quality tools and will perform well for many years. A vise's function is simple: Hold your hook securely. Once you have selected one that will do just that, the tough work is out of the way and it's time to get your boxes filled with all of the flies coming off this new vise!

BOBBINS

Talk about an area of fly tying that truly gives us choices, from varying colors, lengths, inserts, tension adjustments . . . the list goes on! A key for me is that the bobbin rests well in hand, and some newer ones with improved ergonomics are a welcome addition. Easily the most critical component is the bobbin's ability to smoothly allow thread to release, all while maintaining a constant tension. Knowing this, I tend to select those that give the tier the ability to adjust applied tension, while also allowing for quick-change capabilities to other spools of thread. A final consideration is that I prefer to spin my bobbins to cord/uncord thread, so I use those having equal weight distribution to both sides. Other thoughts and recommendations include:

- I prefer a ceramic insert in my bobbins, which prevents fraying as the thread exits the tube.
- Bobbins tend to be used primarily for threads, but I also use them for the easy application of

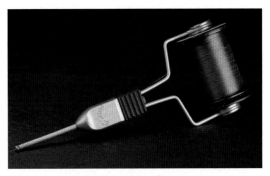

The Stonfo Bobtec Bobbin features a center tension adjustment on the stem; having it there makes changing spools a simple and straightforward process.

The Rite Bobbin (bottom) and Stonfo Elite Bobbin (top) both feature drag systems that come into play when putting thread pressure on the fly. Changing spools requires the side adjustment to be unscrewed, which adds additional time to the process. This is less critical if you tend to select only a few thread colors while tying.

wires, thinner yarns, and other materials that stay spooled with lighter tension.

- Many manufacturers offer bobbins in varying lengths, and my preference is the standard size. However, many tiers prefer a shorter tube, which can give the tier more control, especially with smaller patterns.

- My "bobbin collection" is absolutely unneeded, yet it's nice to have various thread sizes and colors pre-spooled and ready to go. Nearly all I own do their job well, yet I find myself consistently reaching for two: the Stonfo Bobtec and the Rite Bobbin. Both are excellent tension bobbins, with the Stonfo gaining an edge as it spins well when cording thread.

- If you're new to fly tying, one of the most confusing parts of using a bobbin is getting the thread through the tube, and a recommended tool is the bobbin threader. Simple to use, the threader gets inserted through the tube toward the bobbin, then a piece of thread is placed within the threading wire. Pull the threader back through the tube and you're ready to tie. When I was growing up in Pennsylvania, my early tying instructors taught me a method that simply involves sucking the thread through the bobbin tube George Harvey style, though using a threader is a more reliable option—especially when needing to thread the bobbin in the middle of a packed tying demo at the International Fly Tying Symposium!

SCISSORS

Look closely in the main drawer of my fly-tying desk and you'll find my first pair of tying scissors. They are in there to simply reminisce over, as modern versions are far superior in most applications. There are lots of considerations to keep in mind, especially knowing the kinds of materials you're cutting. To work with the different material types, I use multiple pairs of scissors. This helps to ensure the longevity of each set. It is most important to select a pair of scissors that rests comfortably in your hand, as they are a frequently used tying tool. Here is a brief overview to give you some ideas.

- Fine tips: When tying trout flies, especially smaller than #14, I prefer fine-tipped scissors that come to a point, like an arrow. Fine tips allow for maneuverability when cutting with precision close to other materials. Be careful using scissors with delicate ends, as they can easily be damaged when dropped. A new favorite pair are the FlyTier SuperCut Straight from Renomed, a company out of Poland producing quality scissors for many applications. Another go-to pair I use are from Dr. Slick, the Razor series, and their "All Purpose" model is preferred for most common cutting applications.

- Serrated blades: The point (pun intended) of serrations is to help "grab" material during the cut, and this is a debated topic in fly tying. Many argue they're unneeded, as serrations typically aren't necessary if blades are extremely sharp. Knowing the latter can be taken for granted, I tend to recommend serrated blades for others, as it's a preference to use them myself. Cutting coarse fibers, or even metal wires, is something I recommend doing with serrated scissors. The serrations can be on one blade or both, and I prefer the latter for heavy-duty applications.

- Straight versus curved. Having used both over the years, I now prefer straight for all fly-tying applications. There were times it "seemed" the curved pair could cut in more difficult locations, but I rarely have a need for a curved edge that can't be cut with a straight pair. Does that mean there is not a use for them on your bench? Not by any means, as I still own a pair of curved scissors that meet a major requirement: They are very comfortable in my hands, so still see occasional use.

- Razor blades. Knowing many tiers turn to these is great, as they are beneficial for more than just deer hair applications. Fly Fishing Show creator Chuck Furimsky recently was doing a fly-tying demonstration and shared with me that he preferred the sharpness for certain purposes, such as cutting large-diameter and GSP threads. Many use double-edge razor blades for "grooming" deer hair after spinning, preferring flexible blades with their ability to maneuver well and round the overall profile of the pattern.

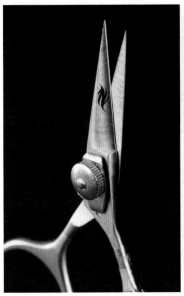 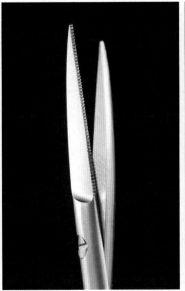 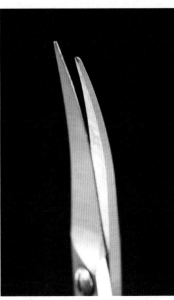

Note the fine tips, which come to a pointed tip, and yes, they are as sharp as they look. Also note that this pair features a tension adjustment, which allows you to tighten or loosen based on the material (left). Note the serrated blade of the Renomed scissors, which have turned into my favorite pair recently. Aside from being lightweight, the scissors rest well in my hand, which is an important consideration when selecting a pair of scissors (center). Note how the tips of these scissors feature a gradual curve. This can be useful when trimming materials into a round shape, such as baitfish profiles or egg patterns (right).

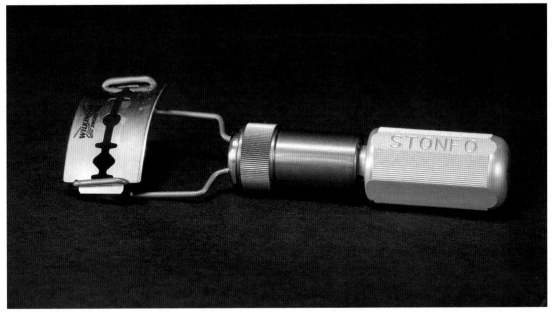

The Stonfo Razor Blade Holder is used primarily for trimming and cutting deer hair. The blade angle is adjustable, controlled with the ring nut, thus keeping the blade safely out of your hand.

One product recently on the market for this need is the Stonfo Razor Blade Holder, which works well, though many tiers prefer to have closer contact by simply holding the flexible blade in their hand. For beginner tiers, I recommend keeping a razor blade handy and using it to remove all materials from a hook when the pattern doesn't come out to your liking.

- Craft store scissors. Don't knock 'em until you've tried 'em! Many tiers around the country use scissors missing the "fly tying" label and have great success with them. A popular pair that my friends use is Fiskars, and there are many brands and models to meet a variety of applications, especially when tying larger patterns and cutting heavy-duty materials.

WHIP FINISHER AND HALF-HITCH TOOL

When teaching fly-tying classes for beginners, one of the first steps I teach is how to finish the fly, as few things are more disheartening in tying than watching a completed fly unwrap before your eyes! Securing materials is critical, and I encourage you to become proficient at doing so. All the patterns in this text were secured with a whip finish, plus additional adhesives and techniques to prevent fish and water hazards from dismantling them. Maybe I'm not the person to talk to for finishing tools, as I use a hand whip finish for 99 percent of my flies. Knowing that many opt for a tool, it's important to discuss finishing techniques for fly tying.

Half-Hitch

This simple knot is used by many tiers throughout the creation of a pattern, and multiple times to finish a fly once tied. Frequently found at the base of other tools, especially various bodkins, using this knot is a preferred method when ease is a concern, notably for beginner tiers. A word of caution: Don't go crazy with this knot! Many tiers add a half-hitch (or more) after securing every material, and the thread buildup can distort the overall proportions of your pattern. Instead, use the knot after tying difficult sections of the fly, then again when finishing your pattern.

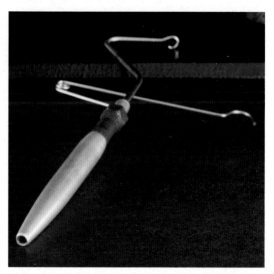

This whip finisher also features a half-hitch tool hidden in its base. If your family is like mine, you may have a drawer full of pens; removing the tip and ink chamber leaves you with a DIY half-hitch tool.

Whip Finish

A preferred method for most tiers, there are many ways to accomplish this, both with and without a tool. With the latter, I have greater control of thread placement and can be precise with its location. My preference is five turns to finish the fly, then pull everything snug before snipping the tag close to the edge. *Tying tip:* Be sure to uncord the thread before completing, which will help to prevent your thread from twisting on itself. Maintain tension until the loop is fully closed, otherwise the thread can twist . . . and snap!

Adhesives

Traditionally, I was encouraged to place head cement on the thread wraps after finishing a fly (nymphs and streamers *only*, never a dry fly!). Since those early days, I've continued to do so, but have expanded this a bit.

- Prior to completing the whip finish turns, I use a light brush to apply adhesive directly to the thread closest to the fly's head. Many adhesives, especially superglues, are sold in a brushable container, which makes for easy application onto your thread.

- Aside from marketed fly-tying cements, popular adhesives I use include Sally Hansen's Hard as Nails. They also offer a "Mega Shine" version, which creates a sheen when applied in multiple coats onto the top of a finished head.
- Superglue is a favorite and used on the majority of patterns in my fly box. You'll notice that the Mop fly techniques in this book use superglue to secure the material to the thread base and shank, whereas the Pliva Shuttlecock has an entire layer of superglue applied directly to the outside of the fly's body. I know one professional tier who simply applies superglue to the tying thread when completing a pattern, makes 15 turns, then cuts the thread (sans whip finish). Traditional? Not by any means. Secure? Absolutely!
- UV resin is a quick way to cover exposed thread wraps; securing with a UV light makes a near bulletproof finish. This technique is a great one to use over fragile materials, too. In this book, there are instances where UV resin is used to secure a material to the shank, cover an entire

Sally Hansen's "Mega Shine" was recommended to me by Greg Heffner, and this top-coat cement features a brush, which allows you to apply the adhesive directly to the thread prior to a whip finish. Multiple coats of "Mega Shine" leave a brilliant sheen on the head of a fly, which gives wet flies that classic look.

abdomen, and keep adhesive eyes attached to a baitfish head. Yes, you can also use UV resin to secure your thread after whip finishing.

FLY TYING LIGHTS

No fly-tying bench can be complete without some form of auxiliary lighting, especially when tying flies in the middle of the night before the next day's trip! With many lights available to us today, don't feel pushed to purchase one marketed directly toward fly tying. Instead, here are some key features I look for when selecting a tying light:

- LEDs: This style of lighting is preferred, as the bulbs tend to last for a significant period of time and give off less heat when turned on. Newer options include models that allow you to control the brightness level, which is an added benefit depending on the time of day. Those that feature multiple LEDs within the lamp are excellent, as they have the ability to diffuse light a bit more over the tying area.
- Adjustments: My favorite tying light has two heads, both of which can be adjusted completely around my vise. This helps immensely with shadows and when I want to highlight something specific. I like having that flexibility, as there are times when I want to bring the light closer to highlight a specific section of the pattern. Many stationary lights are excellent but limit your ability to move them, either around the tying area or to another location.
- Beam pattern: The lights I tend to select have more of a spotlight beam pattern versus flooding light over the entire area. This is a fine balance, as I look for lights that illuminate a section of my fly and vise, all while doing so with a concentrated and bright beam. This is the direction I recommend to others when lighting is also available overhead, from outside, or from an alternative light source. Try to check the beam first because when one is too concentrated, only a small portion of the fly or tying area becomes lit.
- Magnification: This is the perfect place to discuss magnification, as many lights have a magnifier built in. For instance, the LED light I purchased for my father-in-law had two flexible

arms, one loaded with LED lights and the other a magnifier. There are also lights available with the magnifier in the center and the light bulb around the outside. These are great options when occasionally requiring magnification for detail work; however, if eyesight is a concern, a pair of reading glasses (commonly referred to as cheaters) might help.

Tier II: Recommended

As we move into other tools, keep in mind that some are more necessary than others, with that need depending on the individual tier. Are all of these absolutely required to tie flies? Of course not, but they will make the process easier at times, potentially speeding up your tying and overall consistency. The conversation will focus on tools that are common and used by many others, with my own experiences driving everything along.

HACKLE PLIERS

Holding on to hackle may never have been easier, especially with the *crazy long* saddles being produced today! However, hackle pliers hold more than simply their namesake, so there's lots to consider when selecting a pair. Turning to these for shorter and even delicate fibers, the most important part is that the pliers hold a variety of materials securely, and I base my decisions around that notion. If you're in the market for a new pair (or your first), here are some thoughts to consider:

- Long- versus short-handled. Being close to the fly is a positive for my tying, though maneuvering around certain materials on larger patterns would be a reason to get a more significant handle.
- Buffer (aka shock absorber). Steady movements are preferred in tying, but that's easier said than done! Clips with a buffer, such as a tiny spring, work to offer a little give when wrapping materials. There are few worse things than filming a YouTube video and having a slight movement that causes a fragile material like condor substitute to break mid-wrap. This style of plier helps tremendously . . . as does video editing.

- Tension versus clip grip. Pliers tend to have a set of jaws that push together, holding material under tension between the two. A second common version of pliers uses a hook within a tube, with the hook grabbing onto the tying material and holding it in place (i.e., the clip grip). These systems hold materials under varying amounts of tension, and remember that extreme pressure is not always desirable. For instance, when wrapping hackle, if a sudden movement is made using a tool with significant tension at the jaws, the hackle may tear from the fly versus slipping out of the pliers, or even break where the hackle stem is secured by the tips.
- Rotation. Maintaining constant pressure is ideal! Many pairs allow rotation of the shaft, which is perfect for keeping the head in a set position under pressure. As with other tying tools, being closer allows you to work with a smaller range of flies, but longer pairs will provide greater stability.
- Materials. Note the jaws of pliers, which may be metal, plastic, or coated with a piece of rubber. I've used many types and tend to opt for metal or a combination of materials, as they tend to hold more securely at the tips with no slipping. There are many fly-tying materials to choose from, so experiment with some of your frequently used materials to determine which pair of pliers provides the right amount of pressure and holding ability. Having multiple types of hackle pliers is ideal for varying situations.

With so many excellent options in today's tying market, this is by no means a comprehensive list, but a few of my favorite pliers that I recommend to others:

- Stonfo Pinza Elite. Absolute "go-to" pair that seems to securely hold *every* material I use, this tool has been a fixture on my desk for many years.
- Griffin Rotating Hackle Pliers. The longer handle and rotation allows access from all angles and works well when you have adequate access around a fly.
- Stonfo Spring Hackle Pliers. Available in short and long sizes, these are exceptional when working with delicate fibers such as biots and quills.

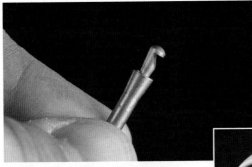

Once the bushing is pulled down on the Pinza Elite hackle plier, you get a peek at how this system works. Releasing the spring-loaded bushing compresses the material, which gets locked in place under the notch.

The Pinza Elite hackle plier comes in two sizes, and this is the standard. It fits comfortably over a finger and takes up little space on your tying bench.

The spring in this set of pliers is intended to offer protection, similar to a shock absorber. Even with an occasional jerky movement, this set allows you to maintain controlled pressure on the material without breaking it off.

- Tyflyz Toolz Hackle Tweezer. An oversized grip makes this comfortable to hold in hand, and a smart design keeps materials securely in place.

DUBBING LOOP TOOL

Dubbing loops are an important part of my tying arsenal, as they can be used for creating buggy bodies, spinning CDC fibers, and more. With a variety of dubbing loop tools on the market (plus DIY options), you have a range of styles and price points to choose from. The ends typically consist of either a shepherd's hook or wire prongs; my preference is the latter, as the tension controls the materials once in place. For many tiers, the dubbing loop tool selection comes down to personal preference and ease of use, and you have lots of excellent options. A few types are briefly discussed:

- Weighted spinner. Convenient to use, these small tools make tight dubbing loops and the weight helps generate many rotations quickly.

No frills on the Dubbing Whirl, yet it offers enough weight to make many rotations quickly. The opening in the wire on top allows you to quickly insert thread in to make a dubbing loop.

Models include the Dr. Slick Dubbing Twister and Performance Flies Dubbing Whirl.

- Stick tools. This type of tool offers more overall control, with the long handle allowing the tier to easily tighten and loosen the loop when inserting materials inside. Popular models include the Kelly Galloup TieUp Dubbing Tool and SWISSCDC Multiloop.
- Ball-bearing models. Combining the best of both worlds, these models spin easily, due to ball bearings found within the head. Being

Note in this picture how the Stonfo Rotodubbing Elite has a head that can be adjusted, which allows you to hold the handle parallel to the hook shank when wrapping on your loop.

housed on a long handle gives the tier additional control, and some models even offer rotation of the head to make wrapping the loop around the hook smoother yet. My personal favorite is the Stonfo Elite Rotodubbing Twister, which is considered a premium tool in this category.

CDC TOOL

It's easy to understand this unique fiber's prominence in fly tying and fishing. CDC has distinct advantages, as the fibers offer tremendous flotation and movement in the water. When you're in a hurry, a quick method includes tying the feather in by the tip and wrapping it like a soft hackle. Since many of the feathers are shorter, hackle pliers come in handy for the task. Other techniques involve removing the fibers from the stem, and there are specific tools available to make working with this much easier, with a preferred outcome of a wing, abdomen, or thorax made of CDC fibers.

Advanced methods involve placing the fibers in a dubbing loop then spinning; using this method for the finicky fibers can have a bit of a learning curve. Now I know what you're thinking: Is it possible to insert the CDC fibers into a dubbing loop, then cut the stem away? Absolutely, but what fun would that be? This paves for the way for fly-tying

tools, and we have many to choose from. The gist of most is simple: Hold the delicate fibers, allowing us to cut away the stems and insert the remaining material into the loop. Here are a few of the CDC tools that I've used over the years:

- Magic Tool. Master tier Marc Petitjean developed this system to encompass many types of materials, with one piece used to stack everything together and the other used as a clamp for inserting everything into the dubbing loop. Overall, the design of this system is intelligent, and there are a variety of sizes to choose from.
- Stonfo Dubbing Loop Clips. Sold in two different sizes, you can determine which would be most useful for your typical tying. The smaller set is recommended for trout patterns, with the larger set beneficial when needing a longer section of spun material. The clips are lightweight and easy to use and are ideal for other tying materials.
- SWISSCDC Multiclamp. Taking the fly-tying world by storm in 2020, this metal tool is longer than most and has the ability to hold a larger section of materials. Made of metal, the clip is the heaviest of this list.
- Loon D-Loop Tweezer. With a longer handle, this tool is ideal for a variety of feathers and

materials and is excellent for creating lengthy dubbing loops for larger patterns.

- Bulldog clip. Sold at office supply stores, this tool has been used by tiers for years and is a great option if you're thinking about trying this tying technique. Some bulldog clips are under extreme tension and can crush materials, but overall, they tend to be the most reasonably priced tool in this set.

THREAD SPLITTER

Another advanced technique is splitting thread, which takes the place of a dubbing loop tool. A brief word of caution: not all threads are created equal, making it much easier to split some than others. To achieve this technique, the thread must be uncorded, which causes it to flatten. From there a sharp point is inserted into the thread, splitting it into two sections. Material is placed between, then the bobbin is spun, creating a quick material noodle of CDC, dubbing, feathers, hair . . . basically whatever you can think of! Some quick tools exist to help with this process, including:

- Bodkin. This tool probably already resides at your desk and can be simply used to split the thread neatly in two. The bodkin is a tool that can be taken for granted, as its simple needle is used to clean hook eyes, free trapped hackle fibers, tease out dubbing, apply adhesive, separate feathers, split thread, and more!
- Needle. No bodkin yet? Try a needle, and I recommend one that is very fine.
- Thread splitter. Created by Stonfo, this tool has a groove where the thread rests. From there, a spring-loaded section gets pushed up, creating a separation in your tying thread.
- Dubbing needle: Another Marc Petitjean design, this tool has a short handle, giving the tier great control to help with this process.

Tier III: Tying Tools Hiding in Plain Sight

There is another tier of tools found on my tying bench, and these tend to be everyday items that I've deemed important and use enough that they warrant mentioning. These are by no means

essential, but most of us have tools on our bench that aren't sold in a fly shop . . . and that's not a bad thing. Continue to experiment and find tools to make tying easier and more enjoyable; basically, don't feel guilty if you're into gadgets, too!

BINGO MAGNETIC WAND

First shared with me by my great-uncle John, this simple tool has a handle and is meant to quickly clear a bingo card. On my desk, it's something I immediately turn to when a hook falls to the floor. Additionally, being magnetized, the Bingo Wand can find a home in many spots, and if I'm just tying a few flies, I'll allow them to attach to this tool until ready for their spot in a fly box.

When the bushing is depressed, a needle is exposed, which will neatly divide your thread in two.

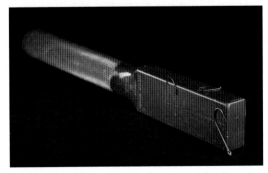

The Magnetic Bingo Wand may not be found at many fly shops, yet this is an indispensable tool on my bench. A quick swipe of the floor with this wand finds a missing hook in no time, preventing that hook from ending up in your shoe or foot.

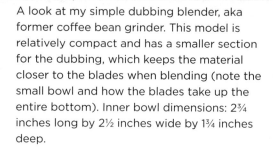

A look at my simple dubbing blender, aka former coffee bean grinder. This model is relatively compact and has a smaller section for the dubbing, which keeps the material closer to the blades when blending (note the small bowl and how the blades take up the entire bottom). Inner bowl dimensions: 2¾ inches long by 2½ inches wide by 1¾ inches deep.

VELCRO

My inner DIY self loves to use the "hook" (aka rough) side as a dubbing brush to make buggier bodies, free trapped fibers, and comb flies into their final shape. Sure, I have tying-specific tools for this, too, like the Stonfo Pettine Comb and Brush, but making your own requires only a piece of Velcro and a popsicle stick—simple enough! On the water, brushing out a New Zealand strike indicator with this tool improves flotability, too.

COFFEE BEAN GRINDER

Don't tell Heather, but our old coffee bean grinder didn't die . . . it was evacuated to my fly-tying room! From it comes a variety of homemade dubbing blends, and each is formulated to look just a little different than yours. The process is relatively simple: I take prepacked packets of dubbing and blend them with others. Typically, I do this to add a minimal amount of flash, though there are times that I'll attempt to lighten or darken a color that closely mimics a natural insect found on a recently fished waterway.

When blending, I prefer to start with a base blend, such as Hare's Ear, then begin to incorporate other colors or flash into it. A word of caution: I have purchased many dubbing packs from the same supplier, all marked with the same blend . . . yet there is slight color variation in all the packs. Knowing this, you may prefer to dye materials if you're more concerned with consistent precision. There are many other techniques to blend dubbing, including with water and air, though the coffee bean grinder is a favorite that has been part of my arsenal for over two decades. Try these tips when blending dubbing yourself (more are included in the section on the Beach Body Stone):

- Between a blender and coffee bean grinder, I've had more success with the latter. My model has an open top where the dubbing is inserted, then can be pulled out of the same basin once blended. The blender that I used had a large container and seemed to swirl the dubbing versus combining the various blends.
- Measure consistently, as this will help you to replicate favorite blends repeatedly. My system

is by no means precise, as my quantities are considered a "pinch" each time one is added. My "pinch" is different than yours, but this allows me to determine the overall approximate percentages of a blend.

- Begin by starting small and adding different types of flash and colors to see how the blend changes. As you begin to find blends that you find yourself using more, scale up the amounts you add.
- Document everything! During each blending session, I keep a notebook nearby so I can tally the number of "pinches" added. Once I reach a blend that has potential, I'll record everything and break it into approximate percentages. Here's an example of a Walt's Worm blend: six pinches (60 percent) tan Hare's Ear as a base; three pinches (30 percent) brown Spikey Squirrel added to slightly darken; and one pinch (10 percent) purple Krystal Flash to add the perfect amount of sparkle!
- Every blend gets stored in a sandwich or snack Ziploc bag, with its ingredients and percentages marked on the outside. As a blend starts to run low, I'm able to quickly see what dubbings I used and re-create it in a short amount of time. Keeping some of the original blend also allows a quick comparison; place one on top of the other, and if there are no discernable differences, success has been achieved.
- *Always* unplug the grinder prior to removing your blended dubbing. Those fly-tying fingers are needed at your bench and on the water, and caution should be exercised when using this tool.

POST-IT NOTES

When tying, my mind tends to clear quickly, and sometimes ideas for tying, fly fishing, or even YouTube pop into it. Jotting them down on a nearby note guarantees their survival, yet I also find functional uses, too. When using UV resins, sometimes an excessive amount is placed on the hook, and I'll use a note to dab away some of the liquid before clearing. As the great tier Tim Flagler has shared in a YouTube video, placing materials between the sheets keeps them elevated and ready

for use, while the sticky backside is great for cleaning up loose fibers. Keep a set of notes handy on your tying bench, and I'm sure you'll find plenty more uses!

NAIL FILE/HOOK HONE

Call me old school, but I still sharpen hooks with dull points. This happens more often on the water, but also at the bench when tying flies for reuse (if only those darn Squirmy Wormies lasted longer!). When I was a young mentee during my teenage years, the fly-fisher mentors I idolized gave me many great tips, and one included using a fingernail file to sharpen smaller hooks. The preferred type is double-sided, with a stainless-steel body; the cost is minimal. There are also fly-fishing-specific tools for this task, though the nail file allows you to easily access all sides of the hook point.

Wrapping up this list of tying tools, let's not forget that there are many more you may undoubtedly deem indispensable, especially considering all the niches within fly tying. A quick examination on and within my bench proves this true, as you'll find hair stackers (and packers), hackle gauges, pliers, foam cutters, wing burners, magnifiers, material clips, UV lights, and more. Each tool serves a specific purpose or fills a need, with the end goal to make tying a tad easier and more enjoyable. We're under no obligation to own them all . . . but some of us may just try!

Materials

Now on to the good stuff: *materials!* Step into my fly-tying room and you'll think you've entered a small fly shop; it houses enough tools and materials to outfit a small country, plus enough flies to fish for the next decade. Some will argue that tying your own flies is a way to save money, and maybe it works for them. However, for those tiers like me, this is a passion, and I'm always experimenting with new (and old) materials; more to the point, there's no money-saving going on! Instead, I prefer to let creativity occur and am constantly finding ways to improve patterns and discover new ones.

SYNTHETICS VERSUS NATURALS

Some premium trout fly-tying items 30 years ago included quality hackle, deer and elk hair, CDC, bucktails, and peacock feathers. Those remain the same today, and manufacturers continue to try to replicate their qualities, with limited success. Thinking about most patterns in this book, there is a mix of synthetics and naturals, with each having a specific reason for their selection. As the sport has progressed over these last few decades, so too has the ability to produce synthetic (and some natural) materials with greater consistency and quality.

There is nothing worse than ordering multiples of the same material online, yet when the items arrive, there are inconsistencies between the two. With natural materials, this is expected to occur, as slight differences occur in nature based on time of year at harvest, individual differences between animals, and even ways in which pieces are cleaned and cured. For many beginner and intermediate tiers, synthetic materials help achieve more consistent results. That consistency is due to improved methods and technology from fly-tying manufacturers around the world. What excites me most is trying new materials and seeing how they improve my tying by saving time and making certain techniques easier. There are many new materials on the market, and let's discuss some.

RESIN ADHESIVES

Scroll through social media today and you will see UV resins in many fly-tying applications. From freshwater to saltwater flies, this is quickly becoming a preferred method for securing materials, especially when factoring in ease of use and speedy cure time. In this book, you'll learn some ways in which I commonly use UV resin, from applying a thin layer to a Perdigon-style body, to securing eyes for a streamer head in the Extreme String Baitfish. As with all chemicals, use resins in a well-ventilated area. Also, I recommend creating a barrier with your hand when curing, which prevents UV light from reflecting into your eyes.

Aside from the few instances shared in this book, think about ways in which you can incorporate UV resin into your tying. Many scud patterns are being shared on social media that have

a resin body that has been built up over multiple applications. Other methods to use UV resin can improve bread-and-butter patterns, such as adding resin around dumbbell eyes to prevent their twisting around a hook on the Clouser Minnow. A less obvious method may be to place a tiny amount at the base of a parachute post to lock the material in place, an easier process versus using thread. Working with delicate quills typically meant counter-wrapping with fine wire; instead, a quick coat of UV resin makes them nearly impenetrable. Another great option is to apply multiple coats to create a bulkier baitfish head; unlike using resins of the past, drying time is reduced to seconds!

Speaking of drying time, a common question I receive relates to drying time with certain resins. Most that I use are cured in under 10 seconds, so when I hear of tiers keeping a UV light on for nearly a minute, I know it's time to do a little investigating. In nearly 99 percent of cases, the culprit remains the same: dying batteries. Be sure to keep fresh batteries in your UV light, as I notice a distinct difference in performance when it's time to change mine.

Many manufacturers have started to create resins of varying colors, some with sparkle included. Creative tying minds have taken off with these

UV resins now come in various thicknesses and colors. Seen here is a small sample of options out there for fly tiers today.

options, integrating them in all types of patterns and styles. When the time came for my own start with UV resins, I was hesitant and overwhelmed (you'll notice there are varying consistencies based on the application). Seeing tier Brad Buzzi at the Fly Fishing Show in Edison, I asked him for a quick tutorial, and he walked me through it . . . then smartly sold me a couple containers of UV resin, which lasted about a month! I was off and running and encourage you to do the same.

THREADS

Centering this discussion around modern materials, there has been an influx of new threads in the last decade, especially those that offer greater tension at a reduced overall diameter. A style that many tiers have been selecting is Gel-Spun Polyethylene threads, commonly referred to as GSP. As with many items on the market, various manufacturers are producing their own versions, most with only slight differences. Concentrating on similarities, here are some quick thoughts and suggestions as you add GSP threads to your arsenal:

- Choose a bobbin that offers a ceramic insert, as other materials can either cut the thread, or worse, some tiers have reported that the heavy-duty GSP can actually put a groove in the bobbin over time.
- Spinning deer hair with GSP threads is ideal, as you can apply a maximum amount of tension while selecting a thread with minimal buildup. The same notion applies when creating a deer or elk hair wing, as the great tier Mike Romanowski used to love tying Comparadun flies with an enormous amount of deer hair for the wing. When he was showing me his method at the International Fly Tying Symposium, I was amazed by the number of hair fibers he was able to lock into place, and asked him his secret. Mike's simple whispered reply, "GSP," said it all.
- GSP threads have a "slippery" feel and can take time getting used to. A suggestion is to consider using wax, which will make it easier to create a dubbing noodle.
- Tying midges with GSP? Go for it, as some companies offer incredibly fine diameters, such as

Semperfli 24/0 Nano Silk. (Yes, you read that correctly!)
- Cutting GSP is difficult depending on its diameter, as it is able to even dull fine tying scissors. Instead, use a pair of craft or fabric scissors, or cut the thread farther in the notch of your all-purpose scissors. Be sure to hold the tag end under tension, then snip close to the hook. This will not be a problem, especially if you are using a quality pair of sharpened scissors. If seeing a thread tag protruding from the fly's head is not your thing, try a sharp razor blade instead.

HACKLE

Talk about a true game changer in fly tying! The quality of today's hackle is nearly unbelievable compared to a short time ago, as the overall length and coloration has progressed substantially in recent years. Yes, I'll admit it: I'm a hackle junkie. The number of flies I tie with hackle has declined in the last few years, as I've turned to using more CDC and deer hair and a variety of synthetic materials in its place. However, set a gorgeous barred dun saddle in front of me and my eyes *always* bug out (right before it gets added to my collection). Understanding the main differences between types of hackle will help you in the long run, especially as you're standing at the "hackle wall" in your local fly shop, trying to decide which one (or five) to buy.

Begin with hackle selection by deciding between a cape or a saddle. A cape's benefit is that it can tie a greater range of sizes; however, the individual feathers will be shorter and there will be fewer overall. A modern saddle's appeal relates to the crazy long feathers, with an individual feather being used to tie up to five or six flies . . . sometimes even more! Most individual feathers on quality saddles may only tie a range of two hook sizes, yet there will be enough material there for two lifetimes. Some quick tips for going about this process:

- Examine individual hackles, ensuring their width is similar for most of the length. Consistent barb length is your best friend!
- Understand the various grades through research. Sometimes it's more beneficial to

purchase a lower grade, which is the same quality as a higher one, but with fewer overall fibers. Trust me, the fish won't mind and your fly will still float proudly on the surface.

- You don't tie tons of dry flies or you're trying to save some money? Purchase a half neck or saddle, or buy the entire piece and split with a friend.

- Base color choices around patterns you typically tie. Favorite personal colors include grizzly, barred medium dun, and barred ginger. Those three colors can get me out of nearly every dry-fly situation, and make a travel tying set easy to pack.

- Unable to examine in person? Call a shop and be specific with your needs. Most of my hackles are intended for dry-fly situations, but I also love their use on various streamer patterns (hence the hot pink and chartreuse feathers hanging out on my bench!). If I called a shop and said simply that I needed hackle to tie streamers, that would leave way too much room for interpretation.

Will it be used Woolly Bugger style, or for tailing options? Plus, what sizes are being tied? I'm positive that Gunnar Brammer's streamers for muskie are just a tad larger than the ones I throw for trout. All this information is critical in finding the right hackle for your tying needs.

Hackle is typically added to my collection based on color and size range, waiting to be called on to fool wise, rising fish. Once purchased, some tiers choose to remove the individual feathers from the skin, then size each one with a hackle gauge. Going this route is time-consuming at the beginning but pays dividends when you want to tie a handful of #16 Sulphur parachute flies and can grab the needed feathers quickly.

If you only occasionally tie flies requiring hackle, consider purchasing packs that include a handful of feathers in a set size. You won't have a lot of money tied up and can purchase for a specific need. Once you've gone through that package, if more is needed you can opt to upgrade to an entire saddle or cape, or an additional package like the first.

In my collection, you'll find saddles in a range of colors, yet that has never stopped me from adding a cape or two, especially in brown grizzly or honey dun.

HOOKS

Improvement in hook selection is easily one of the greatest advances for tiers, in my opinion. The overall structure of fly tying and fishing revolves around this essential component, and landing the fish of a lifetime depends on a quality hook. Kevin Compton, a good friend and the owner of Performance Flies, loves to say that "we can all appreciate a good hook." The "we" aspect is because many of today's hook designs are driven by many groups, including the worlds of competitive fly fishing, warmwater enthusiasts, streamer junkies, and more.

Hook selection may seem straightforward, though I acknowledge that seeing so many brands, styles, and sizes can be a little overwhelming. (We're starting to see a theme in fly tying, right?) Understand that the differences are there to benefit us, as each is intended to fill a niche. A common request I get is to create a hook guide to compare models, though this task would be never-ending, as new brands and models arrive yearly. For most tiers, there are a handful of essential hook models, with the additional ones filling a specific need. Newer models recently released relate to streamer patterns and even certain stages of dry flies, such as their emergence. I'm also starting to see more styles related to Euro nymphing, such as small jig hooks that can hold larger fish. Barbless hooks continue to improve and be offered in a variety of styles, and you'll note that every fly in this book is tied on a barbless hook, as that has been my personal preference for many years.

Does that mean you should select five models and never stray? That sounds easy enough, but pushing to understand more about the impact hooks have on the fly is an important aspect of tying and fly design that shouldn't be taken for granted. An example you'll see illustrated in this book is the use of jig hooks and slotted beads. When used together, the hook is encouraged to ride hook-point up, reducing the number of snags and keeping you in the game for a longer period. However, if the fly is now "upside down," our tying must change accordingly . . . or else your friends may laugh at you on social media!

Quick thoughts on today's hooks:

- Barbless. These options are everywhere and my preferred style for flies today, especially as a proponent of catch-and-release. Once I've landed a fish, the lack of a barb allows me to remove the hook quickly and get that fish back to its natural environment in a short period of time. Being that I own many barbed hooks purchased in the past, I debarb those hooks prior to tying a fly. This simple and quick process involves pliers with fine jaws, preferably flat and smooth inside. In a pinch, you can also use the jaws of your vise to debarb a hook.

- Wide gaps. The gap is the distance between the shank and hook point, with a wider one allowing for more fish to be hooked and landed. In the past, a negative was that some hook styles, especially smaller ones, tended to bend out due to a wide gap. Today's higher-quality materials with a heavier gauge wire can produce much smaller hooks (e.g., #20 jig hook) that can land a large fish without bending out the wide gap. Many styles are now incorporating this, including dry-fly emerger, nymph, and streamer hooks. When I reach for an older pack of hooks, I am amazed at how minuscule the gap used to be . . . and find myself quietly placing them in the bottom of my hook drawer!

- Consistency. Years ago (too many to count!), I would debarb my hooks at the bench and occasionally one would break, not because of too much pressure on my part, but poor tempering by the manufacturer. It was always better to have that occur *before* spending time on a fly that had the potential to break on the water. Common hook imperfections include open eyes, incorrect tempering, and dull points, but today's hooks are much more consistent, with a very high percentage of hooks being "perfect" out of the box.

- Hook points. With many premium hooks, the points appear (and feel) razor sharp and offer slightly upturned tips that keep fish hooked. With certain hook models, a common description others use is that they are "stickier," meaning it is easier to hook fish with the hooks being so sharp and hold on to them through the landing, due in part to longer and upturned points. That's a major positive, but be careful: We have to tie on these hooks, and it's easy to draw blood (of our own) with many styles, especially the larger, articulated streamers.

EXTRA WEIGHT

As fly fishers have constantly battled drag and surface tension to encourage flies to sink faster and deeper, one of our main fighters is the use of weighted beads, eyes, and heads. Aside from weight, some systems help incorporate jigging motion into the pattern, with others using a bead as a hot spot to draw the fish's attention. Each has its own place on today's flies, though they come with advantages and disadvantages. It's easy to shy away and stick with what you know (i.e., dumb-bell eyes—thanks, Mr. Clouser!), but I encourage you to learn more about each type of weight below and determine ways to incorporate the various styles into your everyday tying.

Favorite methods to increase weight:

- Bead head patterns. You'll notice many in this book, as they are easy to tie with and add to the fly's design. Available in a variety of sizes, weights, colors, and even textures, weighted beads are an absolute favorite that help drive my patterns lower in the water column.
- Cone heads. Taking things a step further from beads, this option is great for a slender front to the fly and is a favorite especially for stillwater streamer patterns.
- Baitfish heads. Simple to tie and lock in place, these heads offer a realistic profile and come in a variety of sizes and colors. They are an excellent option when wanting "eyes" on your baitfish imitations.
- Dumbbell eyes. Speaking of eyes, it's tough to beat an original. Many options exist with these today, including color variations, material types, double pupils, 3-D eyes, and more.
- Tungsten bodies. Want to add *tons* of weight to nymph and scud flies? Look no further, as these really will sink your patterns. Keep in mind to use this style of weight in fast water or else these high-priced options will be snagged and lost on the bottom of the water column.

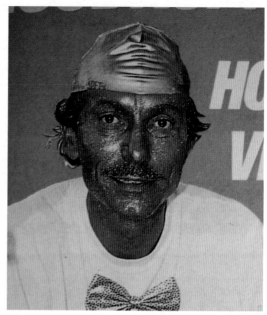

Theo Bakelaar is credited for helping to bring the bead head to the United States. To help popularize the innovation, Theo painted his entire head gold! Photo by Kees Ketting, courtesy of Theo Bakelaar

Does a tenth of a millimeter really matter? *Yes!* Each of these is only 0.1 mm apart, yet it's easy to see the difference, especially when tied onto a fly. Use that to your advantage. I prefer fine diameters more for protection of materials, to create a slight color contrast, or even to encourage the abdomen of a dry fly to sit lower in the surface film. As the wire's size increases, the weight does too; thus you'll find larger-diameter wire on many of my larger nymph, stonefly, and streamer patterns, helping to keep the fly at a deeper water level.

- Sculpin helmets. One of my favorite options when fishing in Pennsylvania, these provide lots of weight and have a keel to help the fly stay hook up. The wide profile at the head is great if there are sculpins or gobies in the water.
- Wire. With a variety of sizes and colors, wire is a great choice for adding supplemental weight to a pattern. Easy methods include lead-free options for underbodies, such as a few wraps of .015 inch wire behind a bead. A wire abdomen in 0.3 mm works wonders in bright colors, as was popularized by John Barr with his Copper John nymph. Don't forget about smaller options that offer weight and protection, especially when counter-ribbing delicate materials. A favorite of mine with emergers is to use a fine 0.1 mm wire to help the body sit just below the surface film.

BEADS, BEADS, BEADS!

An easy approach to adding weight to patterns is the use of beads, popularized in the United States by Theo Bakelaar of the Netherlands. Be aware . . . there are *many* types of beads out there, and I anticipate more to come. That benefits us as tiers, and if you're new to the world of beads, this cheat sheet will help you narrow the choices.

Common materials:
- Tungsten. The heaviest version of beads, this is my preferred choice when I expect the patterns to sink in a hurry. Want to get it down extra fast? Try using two tungsten beads!
- Brass. To me, this is the "original" bead, a weighted version that will sink your patterns to a moderate level.
- Glass. A great choice for emerger flies, as these will allow the fly to sink about six inches below the surface.
- Plastic. These encourage little sinking, and I tend to select this type when simply wanting to add color, some translucency, or a hint of sparkle.

Holes:
- Recessed. With a small circular hole on one side and a larger on the other, this bead is preferred for use on down, up, and straight eye hooks.

Regarding the shank, I love to fish these on a hook with a curve, such as a scud model.
- Slotted. Featuring a circular cut on one side and a slotted one opposing, this is the primary bead I purchase for jig hooks. Though the bead can be finicky when getting them to sit just right, they encourage your flies to ride hook-point up . . . more fish and fewer snags! For a sneaky approach to use these on other hooks, be sure to pay attention when learning about the Mini Jig Bugger.
- Offset. This inverted style is newer to the scene and is preferred on a straight-shanked hook. The premise is simple: The bead encourages the fly to invert in the water, as the weight is concentrated on one side of the bead.

Finish:
- Colors. Your choices are endless, spanning the spectrum of the rainbow; there are even beads that feature a rainbow of colors! Add to this glow-in-the-dark options, glossy ones, matte, painted, gritty exteriors, and even options with nymph eyes on the side . . . it's impossible to try them all (but some of you may try!).
- Aside from weight, two additional methods when using beads include as an attractant (i.e., hot spot) or to blend in with the pattern. My go-to color choices include silver, copper, and matte black, with hi-vis colors such as fluorescent orange and pink rounding out favorite choices. Other selections are added during various situations, such as with pressured fish or high water.

PEACOCK SUBSTITUTIONS

New products are brought to market on what seems like a daily basis, and a growing trend among manufacturers is to find a suitable, if not perfect, replacement for some of the natural materials we love. Nothing says fly tying more than peacock feathers, and both herl and the stripped quills underneath are favorites for many tiers. The fine fibers dance in the water when wet. The colors found on the herl are iridescent, appearing as a variety of colors depending on the angle you're viewing them from. In short, I can't imagine fly tying without peacock feathers. Common types and uses include:

- Packaged herl: A first purchase for many is a prepacked bag containing strung peacock herl, fibers from the tail section that are also found surrounding the peacock eye. Available in a variety of lengths, herl is commonly used as a body material, wrapped around the shank of the hook to provide an iridescent movement to patterns.
- Peacock eye feather: Examining a peacock in real life, the main event occurs as they lift their tail section, covered with colorful eye feathers. As a tier, this is the jackpot! Barbules can be stripped off the colorful eye quills, creating a naturally segmented fiber that is commonly wound to create slender mayfly and nymph bodies. Each eye feather is surrounded by peacock herl, and I have found this material to be of higher quality than those found prepackaged.
- Peacock swords: Originating from the side tails of a peacock, the barbs come to a natural point and tend to be shorter than typical herl. Tiers use this material for tail and wing sections, tying the swords in by their base.

Thinking through the above section, nearly *every* style of fly in this book has a body part that peacock can be substituted for, or the addition of a peacock fiber will enhance the fly. Does that mean this material is free from deficiencies? Of course not; the material can be delicate and easy for the teeth of fish to quickly destroy. For some tiers, the simple act of wrapping a stripped peacock herl is challenging, so other options are used. In 2014 16 species, including peacock, were added to CITES Appendix III, stating that peacock feathers originating from Pakistan now need an export permit to be imported into the United States (https://cites.org/sites/default/files/notif/E-Notif-2014-014.pdf). Tiers and manufacturers took note, and the push for suitable substitutions increased, and will continue to do so.

In fly tying, substituting materials is commonplace, as typically there are materials that can achieve similar results. Finding a replacement for peacock, on the other hand, is not an easy task, yet here are a few that I've recently found to be more than acceptable, and in some cases, nearly better than the original (sacrilegious, I know!).

- Jan Siman Peacock Dubbing: This has been my "go-to" dubbing over the last few years, as the color range is excellent and the material is easy to dub onto thread. Being a dubbing, this preferred thorax and body material also has some beneficial options for tiers during application. When desiring a tight thorax for a slender-bodied pattern, I use a dubbing noodle that has been compacted tightly to the thread. However, peacock herl with its short fibers is known for its ability to undulate in the water. Using a dubbing spinner with this Siman material encourages the fibers to splay out once wound around the hook shank.
- Semperfli Straggle String Micro Chenille: Recently, the fibers of many chenilles have gotten finer and shorter in overall length, doing a great job to replicate shorter feather barbs and, you guessed it, peacock herl. This one from Semperfli is the finest I've used of the fritz-style chenilles, and the variety of colors currently available gives you many choices for a range of pattern types. Being attached to an inner core makes tying in a cinch; thus this is a material I recommend, especially for those new to fly tying who may have problems keeping herl from breaking while wrapping.
- Synthetic peacock quills: Using peacock quills for dry-fly and nymph bodies is something that continues to grow in popularity; however, the fragility of natural materials has encouraged a market increase for synthetic manufacturers. This is to our gain as tiers! I've used those from Hareline, Hemingway's, and Semperfli with success in my patterns, though keep in mind that there are subtle differences between each. Some companies have various sizes; others offer a tapered strip. A further enhancement is the addition of adhesive backing, which encourages the strip to stay in place during tying. Add to this Perfect Quills, which are transparent with a black border, and you now have the ability to determine the color tied underneath. Similar to the natural materials, I tend to coat these with a UV resin for added protection and overall sheen.

Future Materials

The above is by no means an extensive list of improvements for fly tiers in the 21st century, and even better, it will continue to grow, especially within the synthetic realm. As more products are developed and tried, our creativity will flourish alongside, pushing tying to new levels. Excitement will surely grow as you head to your local fly shop or favorite online retailer to view the new assortments and color ranges of future products, perhaps too many to count! That can be a difficult thing, as there is only so much time to test and determine which is best suited for our personal tying, but I'm sure you're up for the challenge, right? From my standpoint, the decision to select a replacement or substitution is guided by a number of questions, including:

- Does the material offer more consistency for my patterns?
- Will this enhance my current flies?
- Is durability or added protection gained from this change?
- Will this speed up my production or make the tying process easier?
- How will this modification improve the pattern enough to catch more fish?

If most of those questions can be answered in the positive, then I'm absolutely willing to give a new product a try, and I recommend that you do the same. There's also the notion of trying new materials simply because they're new and interest you, so integrating them into some of your previous designs may spark more creativity. Just like you, I look forward to these new materials and know you'll agree as I exclaim, "Challenge accepted!"

Social Media

Moving forward from the present time in fly tying, I truly believe that once many have reflected on this period, the most influential component will be social media. From the ease of instruction offered via YouTube to the creativity seen on an endless basis through Facebook Groups and Instagram, the impact of social media on fly tying is truly hard to measure. Posts are made day and night around the world, with the authors consisting of everyday tiers, professionals, pro team members, manufacturers, fly shops, and ambassadors; all of whom are considered influencers, whether they know it or not. As this information comes at us in an almost frantic pace, we must somehow personally curate as much as possible, sorting those we connect with most from a variety of standpoints, including informational, entertaining, and even artistic. This is no easy task.

Personally, I have run a YouTube channel for nearly a decade, sharing videos on specific patterns, techniques, guest tiers, and even social awareness causes related to fly fishing. As the years have gone by, the channel has grown with me, in a sense chronicling my journey as a tier and fly fisher. I'm proud of my growing subscriber base and 3.5 million views, which to me represent people around the world that I am able to help and inspire. As my social media presence grew, so did the number of events I was invited to as a lead presenter. This direct connection demonstrated the power and positivity of social media, bringing many of us together in ways that were unthinkable 20 years ago.

Does this mean all things social media are perfect? Of course not, nor would I ever want to present it that way. For example, since fly tying has been combined with social media, certain products are seemingly being used by every tier out there, causing me to wonder if I've missed something because I didn't see the need for that product (spoiler alert, I didn't). When you notice this happening, I recommend contacting a few tiers that you trust and ask for their opinions, especially relating to how the material works on the water. Advertising is able to find its way into everything associated with profit-making, and though a tough area to exploit, fly tying is still vulnerable . . . don't we all want to find that perfect and easy-to-tie fly that catches all the fish?

That leaves us as the curators who determine the direction of social media and the type of information we choose to follow. Interests and needs change over time, but here are a few ideas I can offer for using social media to your advantage as a tier:

- **YouTube.** Commonly referred to as the world's second-largest search engine, this social media form tends to be more useful when desiring fly-tying and fishing information related to patterns, techniques, tips, and more. YouTube is a platform for its users to post and share videos viewable to almost anyone in the world. An added benefit is that the information is located on individual channels for a long period of time, and you can save videos within a private playlist, referring to them in the future as needed. For me, two primary methods of use have emerged. First, if there is something specific I'm looking for, I search within YouTube, and more often than not am able to find the needed information. Also, I enjoy following other YouTubers by subscribing to their channels, watching their own journeys through tying and fly fishing. A downside of YouTube is that current trends tend to be slightly behind other forms of social media, but the overall quality and information available makes this a premium method of obtaining useful information.
- **Instagram.** Talk about an addicting platform! Instagram consists of images and videos posted by users, and I am subscribed primarily to fly tiers and fly fishers. Once you've created an account, you are able to subscribe to others and can follow "hashtags," which are ways to sort posts. Are there many "hero shots" of a brown trout held high and proud? Absolutely, but get past those and you'll notice that tiers are sharing the very patterns coming out of their vises, nearly in real time. The biggest takeaway I've had from Instagram is that the typical and accepted "norms" of fly tying are thrown out the window . . . and that's a good thing. Creativity runs rampant, with exploration of techniques and materials at nearly warp speed. Another advantage is the ability to connect and contact others, either by leaving a comment on a specific post or by sending others a direct private message, commonly referred to as a "DM." For me, the greatest downside is that I get trapped looking at way too many unique and dynamic patterns and want to tie them all!

- **Facebook.** This area of social media was one that I resisted, as I thought of it more to connect with family members rather than fly tiers around the world. Boy, was I wrong. Facebook consists of pictures, videos, and text posted by users, and to view these posts, the users must be connected as "friends." This works well enough, but limits you based on the number of friends you currently have added. To provide more value to the Facebook community, an area titled "Groups" was added, and then the fun began! Think of Groups as pages in which people with common interests share information with other group members. For fly tying, there are hundreds of Groups that are public, making access easy. Once within them, users can post for all to see and comments are allowed (and encouraged) on one another's posts. These groups are like modern fly-tying forums, with the one downside is that once information is seen, it is not easily accessible again. The true value in Facebook Groups is connecting with others and being inspired by all the creativity shared on a regular basis.

Nearly two decades into this century, and I believe fly tying has grown tremendously—to the benefit of all of us. Accessible information, ways to connect with others, and material unlike anything used before . . . yes, we are very lucky to be tying today. The learning curve now compared to when I first started tying in 1989 has been considerably lessened and will continue to be as new technologies are developed for this great sport we love. In short, a lot has been accomplished in the last two decades, yet I still can't wait to see what the future of fly tying holds!

DRY
FLIES

Corn-fed Caddis

The Corn-fed Caddis, loaded with CDC, offers lots of movement when floating on the water, plus the ability to suspend a lighter fly beneath.

The world of dry flies tends to be expanding slightly; our traditional "match the hatch" patterns are being exchanged for more suggestive offerings, and the Corn-fed Caddis fits the latter mold. Packed with all kinds of fly-tying goodies (aka techniques), this Lance Egan caddisfly imitation floats well on the water, gives off a buggy appearance and, most importantly, can be seen by us at distance. Add in that trailing shuck, and the pattern can match a lot of stuff on the water. Hey, wait, I thought we were getting *away* from entomology . . . tricked ya!

CDC is a critical material for this fly and many others. I am encouraging you to either stock up when it's on sale or become friends with a duck hunter. Notice that we've selected three feathers for

Corn-fed Caddis

- **HOOK:** #14 Hanak 130 BL
- **THREAD:** Tan Semperfli 12/0 Waxed
- **TRAILING SHUCK:** Silver Tan Super Gotcha Ice Fur
- **BODY:** Bleached ginger Awesome 'Possum dubbing
- **RIBBING:** Tan Semperfli 12/0 Waxed
- **WING:** Natural dun CDC feathers (tips)
- **WING SIGHTER:** Chartreuse Parachute Post material
- **LEGS:** Natural dun CDC fibers

the wing alone, and another for the legs. During my YouTube videos, you'll consistently hear me remind viewers to maintain slender bodies for many styles, and I encourage you to do that for most dry flies, but when wanting a CDC wing for a pattern that can be used as a dry-dropper, pile on those feathers! Providing excellent movement when wet, the fibers also offer buoyancy, both for the pattern and its potential use in a dry-dropper setup (more on that later).

Let's break this fly down into digestible pieces, all of which can be applied to other patterns at the bench. A trailing shuck is an important component discussed in detail with the Condor Emerger, and you'll find it on many flies in my box. Like mayflies, caddisflies go through a life cycle, with a vulnerable stage being emergence. During this pupa-to-adult transition, the pupal skin gets "shed," though it may remain attached for some time. The trailing shuck mimics this, while also alerting the fish that they have an easy meal waiting for them, as the caddisfly adult is attached to its skin, awaiting the first flight (hopefully away from a hungry trout).

Continuing to the ribbing, thread is simple enough, intended to provide an extra layer of durability to the body. Another option is to use a light tippet, such as 6X, which will be even more resilient than the thread. As you move toward wire ribbing, the potential for the fly to sink tends to outweigh (pun intended) its durability. Regardless of material, the ribbing helps pull everything together, ensuring a slender and protected body.

Once the CDC wing is locked in place, the pattern has the feel of nearing completion, but the parachute post material makes a welcome addition for seeing the Corn-fed Caddis on the water. Whether it's in low-light situations or fishing at distance, anything that helps you pinpoint the pattern once it has landed is a win in my book. Colors such as hi-vis pink, orange, and chartreuse are prevalent in my emerger and dry-fly boxes, finding their way onto parasol and parachute posts, midges, and even spinner wings.

Working with CDC can be a tough task, but specialized clips have been created to hold the delicate feather (and other materials). At this point, the CDC can be trimmed along the stem, then the fibers will be inserted into a dubbing loop.

A final technique is a dubbing loop for the CDC fibers. This takes more time than simply wrapping the fibers like a hackle because they are no longer attached to the stem. That time is worth it, as I feel they have increased movement and really seem to wander all over when on the water's surface. Try this method for other emergers that you have had success with, especially caddisfly imitations. All things considered, the Corn-fed Caddis is loaded with lots of nuggets, and I encourage you to integrate many into your daily tying.

✓ Tying Tip

Prior to spinning CDC, control its length based on where you trim the fibers off the stem; you also have placement flexibility within the dubbing loop. Understanding this allows you to use a standard-sized CDC feather for a range of patterns, another reason why I turn to them for many dry-fly and nymph patterns. Experiment with the length of fiber that works for your typical patterns, and it may slowly replace many materials, such as Hungarian partridge fibers, deer hair wings, and even traditional hackle.

✓ Featured Technique

Creating a dubbing loop for CDC is something I really enjoy doing, both on nymphs and dry flies. Just like any new technique, getting the steps down has a slight learning curve, but you'll be rewarded in the long run and find yourself incorporating it on a regular basis. Here are some techniques that I've learned along the way:

- There are tons of tools out there to help with this process. For me, a critical one is a clip to hold the CDC fibers. On my bench, you'll find the Stonfo Dubbing Loop Clips, which are ideal because they come in two different sizes. There are lots of other options out there, both inexpensive (simple bulldog-style clip used for paper) and high end, such as the SWISSCDC Multiclamp. Whichever you use, the key is finding a "third hand" for your bench while you make the necessary thread preparations.
- Some threads can be split to create the loop for the CDC; try to separate yours with a needle or bodkin. Once apart, slip the CDC in and spin your bobbin until the fibers have helicoptered symmetrically. Complications I've had included thread that doesn't split easily (frustration #1) or bobbins that spin too wobbly due to uneven weight (frustration #2). Experiment with the threads and bobbins you have, and if you're not having luck, see the next tip.
- There are many tools out there to help create the dubbing loop, and a benefit of these is that they are quick and provide the needed tension

when you're ready. My favorite one is the Stonfo Elite Rotodubbing Twister, as ball bearings allow it to spin smoothly, and the head rotates when you're ready to wind the loop around your hook. Many other dubbing twisters can be found with a quick Google search, so see which one is right for you.

✓ Materials to Consider

With modern materials, you have too many options when ribbing this style of fly. Thread makes for an easy choice, as it's a simple technique to use the tag end. Here are some additional options that should find their way onto your dry flies:

- Tippet: By using 6X on flies, you create pronounced ribbing, and today's tippets have excellent breaking strength.
- Sulky Sliver Metallic: Found at many craft stores, this material comes in a variety of colors, giving the perfect amount of sparkle to your patterns. Sulky is super strong and provides excellent durability.
- Extra fine wires: Wire on a dry fly? Absolutely! Today's fine-diameter wires offer incredible protection to delicate body materials such as biots and thread. A go-to wire for me is Semperfli 0.1 mm Tying Wire. Yes, it's really that tiny!

✓ Fishing Suggestions

European nymphing is the current "rabbit hole" I keep going down, and it has reintroduced me to other fly-fishing techniques, with a notable one being dry-dropper. The gist is simple: A dry fly helps suspend a lighter nymph. The pattern combinations are nearly endless, making this a fun style of fly fishing. When fishing a dry-dropper setup, I would traditionally tie the tippet for the dropper through the eye of the dry fly or onto the bend; neither was ideal, but any movement on the dry sent the alert that something was going on with the dropper below.

Enter advanced techniques, and I now fish this setup on more of a Euro-style leader. When fishing a dry-dropper rig, I like to attach the dry fly to a dropper tag tied to my main leader with a triple surgeon's knot; the dropper is connected to the tippet as the point fly. Advantages of this setup include:

- The two flies remain connected yet independent of each other. Being that the dry is on a dropper tag, it appears more lifelike on the water, moving with the seams yet also going under or spinning when a fish has taken the dropper.
- Traditionally, changing the dry fly meant also having to adjust the dropper; this setup allows you to change one fly without impacting the other.
- When paired with a Euro line or mono leader, you can achieve incredibly long drifts, perfect for dry-fly fishing. Casting with this setup can be tricky at first, but once mastered, you'll find yourself choosing a dry-dropper setup in many situations.

Tying the Corn-fed Caddis

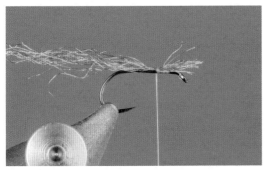

1. Secure the thread one hook eye back, then wrap rearward. Trim the tag end with scissors.

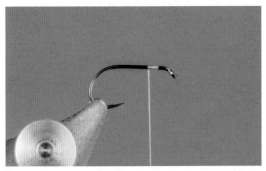

2. Trim a small clump of Ice Fur for the trailing shuck; moisten it to help keep the entire

section together. Secure it with two loose thread wraps, allowing the fibers to protrude over the hook bend.

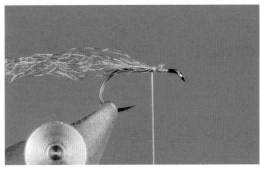

3. Gently pull the Ice Fur butt fibers rearward, moving their ends closer to the thread wraps. Once proficient with the technique, this step can be combined with the prior in a fast manner, saving time because there will be no need to trim tag ends with scissors.

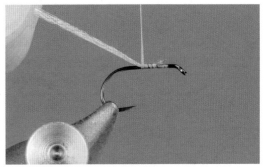

4. Wrap the thread toward the bend to secure the Ice Fur. To ensure the shuck lies on top of the hook shank, hold the material with your pointer finger and thumb, maintaining upward pressure and pulling the material slightly toward your body. Each wrap will encourage the material to move slightly around the shank, lying perfectly down the center.

5. Wrap the thread to the bend of the hook, securing the Ice Fur on the top of the shank. Trim the trailing shuck, keeping its length approximately one-half of the hook shank.

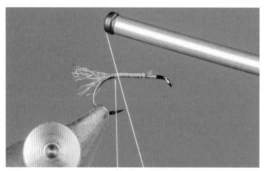

6. Create a loop in the thread by placing your pointer finger below the hook and taking one wrap around the bottom of your finger. After returning the thread to the hook shank, place two wraps in front and behind the thread loop to lock it in place. The overall length of this loop should be no greater than 6 inches.

7. Holding the loop above the hook, trim one piece of the thread close to the hook shank. Be careful not to cut the entire loop off!

8. Secure the entire piece of thread in a material clip, with it going in the direction of the tail. This will eventually be used to rib the abdomen.

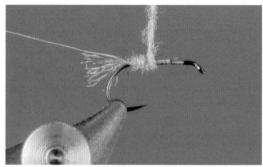

9. Create a tight dubbing noodle on the thread, keeping it sparse on the section closest to the hook. As you create the dubbing noodle, allow more material onto the thread as you get farther from the shank. This will help create a slight taper toward the thorax of the fly. Begin winding forward toward the eye with touching wraps.

10. Continue winding the dubbing noodle forward with touching wraps, allowing for a gradual taper to the body. Once you've reached the thorax, stop winding and remove any excess dubbing from the thread.

11. Remove the thread tag from the material clip and begin to counter-rib toward the eye of the hook. Ensure that the thread wraps are spaced evenly, allowing for a gap between each turn.

12. Continue to counter-rib the thread tag, stopping once you've reached the thorax. Secure the tag with your standing thread, then trim any excess.

13. Select three CDC feathers and straighten them so the tips are aligned. Measure them against the shank, ensuring the tips extend just past the hook bend. Transfer the feathers to your left hand, keeping them aligned against the hook in the position where you want to tie them in.

14. Lock the butt ends of the CDC feathers in place with a few thread wraps, then verify the length of the fibers. Adjust them to the desired position (if needed), then lock in place with additional wraps. *Tying tip:* In this picture, I demonstrate that if the tips are extending past the bend too far, you can coax them into position by gently pulling on the butt ends.

15. Trim the butt ends of the CDC fibers close to the hook shank.

16. Trim a piece of parachute post material approximately 2 inches long, then pinch away a half strand. (Save that half for your next fly!) Loop the parachute post around the thread, holding the ends between your thumb and finger.

17. Wrap the thread around the shank, allowing the parachute post material to move to the top of the fly. Pull the material to the left, encouraging it to sit over the CDC fibers.

18. Continue wrapping the thread around the shank, but intentionally hold the parachute post material in place on top of the hook shank. Make additional thread wraps, locking the post material securely against the hook.

19. Continue with rearward thread wraps toward the wing, securing the post material to the thorax. ***Tying tip:*** To ensure the material remains centered, pull it slightly toward your body while wrapping. The thread wraps will help carry it over the shank, lining it up perfectly with the wing.

20. Trim the parachute post fibers, ensuring they're slightly shorter than the CDC wing. This will make them visible to us, but not to the fish.

21. Create a thread loop, with the steps shown here using a Stonfo Elite Rotodubbing tool. With the thread tight to the left side of the thorax, bring it down into the hooks of the tool in a counterclockwise direction. Start by catching the thread in the left clip, then over to the right one.

24. Remove the thread loop from the tool and place it out of the way; in this picture, the loop has been placed to the left side of the fly, into the vise's material clip.

22. Return the thread to the fly, then continue winding it over the shank. At this point, you'll notice a triangle being formed with the thread loop.

25. Create a thin dubbing noodle on the thread, then advance toward the eye. Be very intentional with the last couple of touching wraps, as it is easy to build up the materials too close to the eye.

23. Wrap the thread toward the left side of the thorax, winding over both sides of the dubbing loop to lock everything in place. Consider taking one wrap on the left side of the loop by bringing the thread underneath, which adds an additional level of security.

26. Place one side of a CDC feather in a material clip. Grip the midsection of the feather, leaving the butt ends of the fibers exposed so you'll be able to trim closely to the feather stem.

27. With a single cut, trim away the CDC fibers from the stem. When cutting, keep the tips of your scissors close to the feather stem, which will help you gather the longer CDC fibers.

28. Remove the thread dubbing loop from the material clip and open the loop. In this picture, I was able to place the loop back into the tool, helping to keep the sides open while I placed the butt ends of the CDC fibers into its opening. Once they are placed in, ensure that the CDC butt ends stay close to the thread loop.

29. Release the CDC from the material clip, ensuring that you are putting tension on the thread loop to close its opening at the same time.

30. Holding on to the clip, let it hang directly below the shank of the hook. Note that the standing thread on the bobbin is to the right of the thread loop.

31. Spin the dubbing loop tool, causing the thread to create a CDC hackle of fibers. The tip sections are much longer than those of the butt. Vary their initial placement based on your preference, though this arrangement gives you a variety of lengths to wrap with.

32. Wind the CDC "hackle" through the thorax with three or four turns, stopping once you've reached the eye. Make the wraps nearly touching, and be sure not to trap any CDC fibers underneath the thread.

33. Secure the loop by the eye with thread wraps, then place an additional wrap in front of the loop to help keep the thread back from the eye. Trim the thread loop close to the hook shank.

34. Whip finish and trim the thread. Be sure to keep the wraps back from the eye, especially when using one that faces down.

35. Examine the thorax of the fly, looking for any trapped fibers. Use a dubbing brush to loosen any CDC that was caught under the thread wraps. Push the brush forward toward the eye, then rearward to the bend, to loosen the fibers.

36. When I tie these, the stragglier, the better! As this pattern moves in the water, those CDC legs will play in the current and give the appearance of life. Vary the body and parachute post colors based on your needs. There is also a dyed hi-vis CDC that can be seen at distance, which will help keep this pattern floating. With all of the CDC used, now you can see why this is perfect when fishing a dry-dropper.

Moodah Poodah

The Moodah Poodah is able to dip its abdomen into the water, encouraging its hot spot and those rubber legs to attract the attention of nearby hungry trout!

In the world of dry flies, many materials help keep patterns floating, and a favorite of mine is foam. Naturally buoyant, this material can keep itself on the surface with little supplemental help, and is a great choice when wanting to suspend dropper patterns underneath. Tying with it can be tricky, so my intention with this step-by-step tutorial is to give you some foam tips and also discuss ways to incorporate visibility into a fly . . . for both you and the fish! Oh, there are some rubber legs to have fun with, too.

Created by Curtis Fry of Fly Fish Food, the Moodah Poodah is a great pattern to pull techniques from, as I consider the fly a base style that can lead to additional variations. Throughout this book, we're learning about differing types of legs,

Moodah Poodah

- **HOOK:** #10 Fulling Mill FM50 65 or 35065
- **THREAD:** Black Semperfli 12/0 Waxed
- **RIB:** Sulky Sliver Metallic (8040)
- **TAG:** Fluorescent hot pink Ice Dub
- **BODY:** Black Fire Flash dubbing
- **WING:** Black deer hair
- **HEAD AND WING:** Black closed-cell 1.5 mm foam
- **LEGS:** Neon red grizzly barred rubber legs (medium)
- **SIGHTER:** Hot pink Parachute Post

dubbing, and now foam, and the Moodah Poodah gives us a taste of some favorite materials already found at most tying benches. Even more, this style of fly incorporates a body that rides in and below the surface film, something fish find irresistible at times. So yes, you're reading this correctly: We've stumbled across the perfect learning pattern that trout also love!

This pattern can seem complex, but let's break it down into bite-sized pieces, starting with a hot spot. The Moodah Poodah's use of one is intentional, yet subtle. A rear tag is something that gets incorporated into many of today's patterns, this one using pink to capture the fish's attention. In some cases, a rear hot spot may represent eggs or a nymphal shuck still attached; in other patterns, it's there to simply be different and create contrast. Find ways to include various hot spots in your patterns, and the fish will reward you for doing so!

The body is straightforward, though I have seen different approaches taken with color variations. My preference is to keep it simple, with darker colors preferred. After the body is finalized, deer hair is chosen as the wing, which helps keep this pattern suspended in the surface film. Only have elk hair? Use it! The material selected should help float the pattern, while providing a silhouette when fish are looking for one. Be sure to stack the fibers and comb out any underfur, as the latter can absorb water and create a bulkier tie-in point.

Time for some foam, which really pulls everything together in this pattern and so many others! Notice that the foam gets bent around and poked through the eye, allowing buoyancy on both the top and bottom of the fly. This is part of the design, which encourages the front of the fly to float higher than the rear. Are you allowed to just tie in the foam on top? Absolutely, you're the tier! Understand that doing so changes the overall appearance, and the entire body and thorax will sit lower in the water. Alter the fly based on your tying and fishing needs, which is what makes this such a perfect learning pattern.

Aside from the foam, another key material is the rubber legs, and I find myself using them on more patterns every year. The reason is simple: Rubber legs are easy to attach, reasonably priced, and

Aside from the variety of colors of foam available today, note the inner composition and how the pieces differ. Matching the right foam to a pattern makes the entire process much smoother.

provide great movement and vibration to attract fish (and tiers). In many flies, the legs are more representational, and seeing them on a dry fly can seem excessive. In my mind, as the fly moves on water, the legs pulse slightly, potentially resembling terrestrial food scrambling in a mad attempt to get back to dry land. Yes, the hi-vis colors may not be a *perfect* match for those insects, but there is something to be said for a hot spot attracting fish.

We finish with parachute post material intended to help us locate the fly when fishing at distance or during low-light situations. Once complete with the whip finish, take a step back and think about all the great techniques that are incorporated in this fly. They are not the easiest, but can be mastered with repeated and intentional practice. More importantly, the techniques and materials used in the Moodah Poodah are translatable to other excellent patterns, so you should have some empty fly boxes ready to fill!

✔ Tying Tip

Like in many things fly tying, not all foam is created equal. With subtle differences that can greatly impact your tying, it's important to understand the nuances involved before buying. For example, some types tear easily when secured with thread, while others require incredible pressure, even with GSP thread. The differences tend to relate to the type of foam that is being used, with your two options being closed or open cell. Let's talk about some points to consider when incorporating foam into your tying.

When sourcing foam over the years in person, I like to remove it from the packaging and check on a few factors, including:

- Softness. Open-cell foam tends to be softer, which is desirable when it comes to trout patterns. Keep in mind, though, that open-cell foam tends to sink, and thus wouldn't be beneficial to dry flies.
- Thickness. Most of my trout patterns tend to use foam 2 mm or thinner, as it's easier to maneuver around the hook and ties in with fewer thread wraps. Thicker foam at my bench gets reserved for larger patterns, especially those targeting larger warmwater and saltwater species.
- Some foams absorb water (open cell) versus those that don't allow air inside (closed cell). We're using the latter with this pattern, with the intention of fishing it as a dry fly. Does that mean you shouldn't purchase open-cell foam? Absolutely not, as it can be used when you want flies to absorb water and ride lower in the water column.

✔ Featured Technique

Let's talk more about foam . . . kinda! During the tying procedures, time is given to carefully cut and measure the foam prior to placing it onto the hook eye. This is an essential part of this pattern and many others, as I prefer to use common hook components to help with the process. For instance, when measuring the width of the foam, my guide is the hook gap. In fact, when I first started tying,

an early favorite pattern was the Woolly Bugger, and I tied these so often that I actually had markings chiseled into my tying bench so I could measure materials once, reducing the number of times I used my scissors! Tiers keep other measuring tools on their bench to ensure pattern uniformity, and two common ones are rulers or flat-leg dividers. By using set individual standards, consistency is maintained with your patterns, ensuring that the end products are similar. Another simple way to ensure consistency is to purchase foam body cutters, as they will help to create identical bodies while also speeding up the cutting process.

✔ Materials to Consider

The Moodah Poodah was an easy choice to consider for this book, as this is the *perfect* pattern to substitute other materials without sacrificing overall effectiveness. This naturally includes changing the color of most components (which you should absolutely try), but let's also examine other recommended substitutions:

- Dubbing. There are various types of dubbing out there, but other great body materials include peacock herl, goose and peacock biots, even thread. Add chenilles and a variety of other synthetics . . . and you have *many* decisions to make! Start with the profile you want to create, then branch out from there. The dubbing used for this pattern has flash in it, but for similar patterns, I've also used natural fur dubbings and enjoy tying bodies with a dubbing that is mixed with micro rubber legs.
- Ribbing. This is used to protect a material underneath or provide prominent contrast from the body. Sulky brand offers subtle flash, and other favorite choices are fine wire and thread. A key with dry flies is to use lighter materials that will help the pattern maintain buoyancy.
- Deer hair. In Pennsylvania, I have an abundant selection of whitetail deer hair, and I use it *all* the time! Go with elk hair if it's in your collection already; if not, a piece of cow or yearling elk is a must-have for similar patterns.
- Post material. The key is to create a spot that you're able to see at distance when fishing. I've

also used hi-vis UV resins on the foam, making it much easier to locate. Another option is contrasting colored Krystal Flash, with a bunch tied in place over the foam.

This style of foam dry fly is intended to represent a multitude of trout food, including hoppers, cicadas, beetles, sunken ants, crickets, and more. If you know the fish are keying on a specific one, try to "match the hatch" accordingly. Some techniques include:

- Fish with force. Some terrestrials land on the water with a loud *plop*, such as beetles and cicadas. In many instances, I'll attempt to drive the fly into the water, with the disturbance causing an instant reaction from fish.
- Drive the middle. Knowing that many terrestrials fall into the water from shore or are forced in by wind, trout are ready and frequently found within inches of shore. When wading or floating during this type of season, I will stay in the middle of the main stem, with my casts reaching crevices along the bank and below hanging vegetation.
- Cast behind . . . on purpose! Once I've located a rising fish tight to the bank on pressured tailwaters, I love to find the perfect emerger. When that doesn't work, I will throw a terrestrial, but by casting *behind* the trout, sometimes 6 inches from the tail. To have a 20-inch brown trout turn and annihilate a #10 Moodah Poodah after throwing tiny mayfly imitations makes for a great change of pace . . . just don't forget to increase your tippet size!

Tying the Moodah Poodah

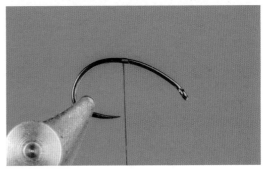

1. Angle the hook eye down when locking it into the vise, as tying will concentrate on the rear part of the hook. Notice how the hook eye is straight across from the tips of the vise jaws, allowing you access down the bend of the shank. Tie in thread, allowing the bobbin to rest in front of the hook point.

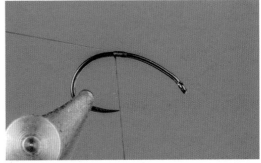

2. Lock in ribbing with thread wraps, then place the ribbing into a material clip.

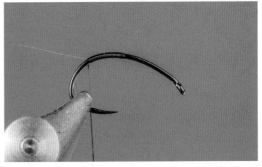

3. Wind thread rearward, past where the barb would be located. At this point, your thread should be slightly down the bend of the hook.

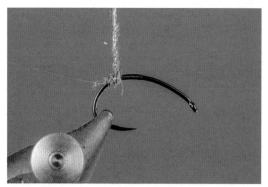

4. Create a tight dubbing noodle, selecting hi-vis dubbing for the hot spot. This one has some shine, which will reflect light from its final position. Begin winding forward with touching wraps.

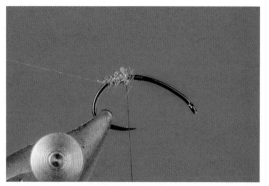

5. Continue winding forward with four touching wraps, then remove any excess dubbing from the thread. I tend to overdub for a hot spot effect, and to ensure I've fully covered the dark thread.

6. Apply dark dubbing to the thread via a dubbing noodle, then wind forward with touching wraps until approximately two eye lengths from the front of the hook. I prefer to select a dubbing color that contrasts with the hot spot, with a peacock blend being a personal favorite.

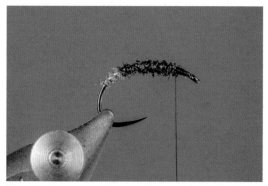

7. Counter-wrap the ribbing through the hot spot and body, allowing for a small gap between each wind. Once meeting the thread, tie off the ribbing material, then cut away excess. Next, wind to the hook eye, creating a thread base, then return to the thorax. Trim any straggly fibers at this point (not the thread!), then reposition the hook in the vise to a standard position.

8. Stack deer hair by the tips, then measure so the tips extend to the hook bend. Pinch the hair in between your left thumb and pointer finger (for right-handed tiers), and lock in place with multiple pinch wraps. I prefer the first wrap light, then increase pressure with each turn of thread. The deer hair butts should face over the hook eye.

9. Wind through the deer hair butt ends, lifting them up with each wrap. This step will fully lock them in place (note how their position has angled up slightly after winding through).

10. Trim deer hair close to the hook shank, then cover the head with thread wraps. Some prefer to use a razor blade to trim the hair, though be careful not to accidentally cut the thread underneath, causing the wing to fall apart.

11. Prepare a piece of foam by trimming its width to match the size of the hook gap. Hold the foam up to the gap to ensure the measurements are close. With many hooks on the market, the gaps will absolutely vary, and the key is to cut the foam wide enough to significantly help the pattern float.

12. With your bodkin (or the tips of scissors), poke a hole in the foam approximately one-third from the end. Place the opening over the eye and slowly slide the foam into position, with the longer section of foam facing up.

13. Fold the foam so that it lies over both the top and bottom of the hook shank, then rotate the vise 90 degrees. Once rotated, the thread will rest through the "sandwich" gap between the two foam sections. Make two loose thread wraps, then pull down firmly on the thread to secure.

14. Rotate the vise back to its starting position, ensuring that the foam is lined up evenly on the top and bottom of the hook shank.

15. Place a set of rubber legs on the side of the hook facing you, locking them in place tightly against the foam. *Tying tip:* If you're having trouble tying in the legs, first get them tight against the fly with one firm wrap, then slide them into the desired position. Once there, make further thread wraps to lock the legs in place.

16. Tie in the second set of legs on the far side of the hook, then trim to the desired length. I tend to oversize the legs, as it's much easier to quickly snip the material off when fly fishing versus finding a way to glue more on!

18. Allow the post material to be pulled down by the bobbin weight, with its resting position in the crease of the foam. Pull firmly on the thread to tighten, then make additional thread wraps to lock the post in position.

17. Fold the parachute post around the thread, then place one thread wrap over the foam. While doing so, hold the post material with your left hand above the center point of the foam.

19. Trim the parachute post so it's approximately the same length as the foam wing. Leaving the post longer is fine, though it may cause the fly to spin slightly when casting. Whip finish the thread in the foam crease, then trim away. If you want a more secure finish, consider brushable superglue or head cement placed onto the thread prior to the whip finish.

20. Examining the fly from the top shows the proper placement of the legs and the length of foam compared to the deer hair wing. Everything is tucked together nicely, and the hook eye is exposed, ready for the tippet to be tied in place.

21. A more important view, this is what the fish see when this fly lands on the water surface. The hot spot is obvious and stands out when contrasted with the body, and the rubber legs are splayed and patiently waiting to vibrate once in the water. The hook is slightly offset in this picture, and even so, few parachute post fibers can be seen from this perspective.

22. Want to get a little fancy? Trim the foam wing by cutting a V into the center, which accentuates the remaining two tips. Will this matter to the fish? Doubtful, but it certainly will grab your attention when thinking about tying the pattern on . . . and impress your friends, *if* you let them look in your fly box!

23. There are many techniques gained from this pattern, though the final one can be viewed from this perspective. Notice how the curved hook allows the hot spot to be seen. Once this pattern is resting on the water, the hot spot will be forced through the film. Designing a fly takes every detail into consideration; use that to your advantage. If you want a more realistic pattern that rides flush, select a straight-shanked hook instead. When varying this style of fly, *many* choices exist, from adding CDC to altering the color scheme. These are patterns I love to tie and fish, and they work well in a dry-dropper setup. Tinker away and see how you can modify the Moodah Poodah to be effective on the waters you haunt.

EMERGERS

Condor Emerger

The Condor Emerger is an imitative mayfly pattern that can be adjusted in color and size to match most mayflies in the waterways you fish. Knowing the nuances of how mayflies emerge give us a bit of an advantage over fish during this vulnerable insect stage.

Over the years, my fly-tying style has changed greatly, related directly to the types of flies that have proved successful on the water. When it came to dry flies, my Uncle John was a major inspiration, sharing with me the intricacies of the Adams and Light Cahill, tied Catskill-style to near perfection. Those patterns still hold a special place in my heart, but my tying transitioned to encompassing more parachute patterns, especially those representing the vulnerable emerger stage of mayflies.

In my late teens and early 20s, I was fortunate to spend time on two special rivers: Montana's Missouri River and New York's Delaware River. Nearly considered "homes away from home" due to the time I spent on each, both offer challenging opportunities for large trout and technical dry-fly fishing;

Condor Emerger

- **HOOK:** #14 Fasna F-120
- **THREAD:** Light cahill UNI 8/0
- **PARACHUTE POST:** Pink Antron
- **TAIL:** Silver Tan Super Gotcha Ice Fur
- **RIBBING:** Brown Semperfli Tying Wire 0.1 mm (fine)
- **BODY:** Ginger Veniard Condor Substitute
- **THORAX:** Pale yellow Delaware River Company dubbing
- **LEGS:** Grizzly dun saddle hackle

throw in pressured water with many exceptional anglers, and you must bring your "A game" to the water daily. These were training grounds for me, and I soon found success learning about one essential style of fly: emergers.

Briefly touching on the life cycle of mayflies: After the nymphs have hatched from their eggs, most tend to forage on the stream bottom before beginning their emergence to the surface. Focusing on mayflies that impact our fly fishing, the nymphs rise through the water column, then rest just below the surface, as pushing through is difficult because of increased tension. It's at this point that the mayflies are extremely vulnerable . . . and fish know it! As a tier and fly fisherman, I've chosen to exploit this stage, hence the Condor Emerger. This specific fly represents the Sulphur, Pale Morning Dun, and Light Cahill mayflies, but by varying the body color, you open the system to a broad variety of mayflies. Let's go through the specific materials; it's my hope that you can apply this style to your own dry flies for increased success on the water.

Let's begin by discussing the pink parachute post, which usually gets a snicker or two from my friends. Knowing that this pattern sits lower in the water, I prefer a parachute post that is highly visible, hence my preferred color choice. Others shades that I employ include fluorescent orange and chartreuse, which greatly aid in spotting this fly during low-light situations. Finally, there can be times when glare is on the water, masking bright colors, so keep a parachute or two with a black post, and you'll be able to spot the fly quicker than with other colors.

Working toward the bend of the hook, notice the use of Ice Fur for the trailing shuck. This is an integral part of the fly, in my opinion, as it is intended to represent the nymphal shuck still attached to dun as it dries its wings and prepares for first flight. Think back to dry-fly fishing where you've experienced rising trout that have eaten duns floating down the river with their upright sailboat wings. The fish eat these happily, but ignore yours . . . why? We commonly blame drag, but I also argue that though it appears the fish are eating duns, those duns may be attached to a nymphal skin riding underwater. This *screams* vulnerability to the fish, leading me to integrate a trailing shuck into the majority of my emerger patterns. Cutting the shuck relatively short helps blend it into the body, signifying that the adult is still attempting to free itself.

Condor substitute is something I was introduced to by Kevin Compton of Performance Flies, an exceptional tier who isn't afraid to experiment with a variety of materials. One of his popular nymph patterns, the Cinnamon Toast Baetis, features condor substitute for the body, and I knew it would be a perfect imitation for nymphal skin getting pushed off the mayfly dun at the water's surface. A key to this material is that the barbules move slightly in the water, appearing as nymph gills. As you'll notice when tying with it, condor substitute is a delicate material; hence a thin wire is used to provide protection while also encouraging the body to sit below the water's surface, similar to the nymph. Few mayflies feature wire ribbing, but this is part of the design for the Condor Emerger.

The thorax has a color shift, intended to represent the mayfly dun extruding itself from the nymphal skin, and this is where I suggest you vary the color based on the natural mayfly duns in the waters you fish. When selecting dubbing, try to use types that allow for tight dubbing noodles, as it's easy for the thorax to become unnaturally thick with material buildup. Some dry-fly dubbing will also have guard hairs, and I prefer to snip them away from the thorax after whip finishing the fly. Before we get that far, let's talk about the parachute hackle.

Condor Substitute is actually a turkey feather, and can be found in a variety of colors. A few of my "must haves" include (from left to right) olive, reddish brown, and pale yellow.

With today's genetic saddles, you can expect to tie many flies from just one hackle. Most have an exceptional number of barbs, and you'll only need a few winds before securing the hackle stem to the post. On many of my YouTube videos featuring parachute flies, I will occasionally finish the hackle by securing it to the body versus the post, as is shown in the step-by-step pictures in this book. A frequent comment I receive related to that style is: "When you secured the hackle, I noticed many fibers are pulled down, and will lie toward the water versus parallel to it. Why is that?" My reply in every case is that those fibers splaying toward the surface appear like an insect stuck in the emerger phase, struggling to break out of its nymphal shuck. That's not a bad thing, as it will encourage the body to lie slightly above, versus in, the surface film. Some of those hackle barbs will push through the surface, adding the effect of life to a parachute pattern. Decide if that's what you're trying to accomplish, and if so, modify the tying to suit your fishing needs. What I'm trying to say is that if your parachute hackle fibers don't appear "perfect," have no fear, the fish will still be obliged to eat.

✔ Tying Tip

After locking the wire ribbing in place, I prefer to use the helicopter technique to tear the tag off. This handy method saves your scissor blades from having to cut metal and will remove the tag end relatively flush with the thread wraps. A key to this method is to make three firm thread wraps to secure the ribbing, then maintain that thread pressure as you helicopter the wire away.

✔ Featured Technique

The final component on this fly is parachute hackle, and there are many methods to executing this technique. My current preference is like Charlie Craven's style, winding the hackle down and securing it with the thread to the parachute post. This technique ensures that the hackle is locked against the Antron, allowing you to next "jump" the thread to the eye versus potentially capturing any hackle fibers underneath an errant whip finish

to the post. Prefer to whip finish against the post? Go for it! Doing so will help keep exposed thread away from the teeth of those hungry trout. Both techniques will enhance your tying of parachute flies, so select the one you're most comfortable with when finishing.

✔ Materials to Consider

We are completely spoiled with hook choices today! Notice that the Fasna has a unique bend to it, its primary purpose being for Klinkhammer-style flies. This shape encourages a portion of the tail and body to sit below the water's surface, exactly what we would expect from an emerger. Take advantage of this hook style versus a traditional straight shank when tying emergers, adding some of the newer options to your arsenal.

Traditional dry fly patterns representing the adult tend to be tied on hooks with a straight shank, encouraging the tail to remain upright coming out of the abdomen. When tying traditional parachute patterns on this style of hook, the hackle can commonly cover the hook eye, making it difficult to thread tippet through when in a hurry or in low-light conditions. An option available to us are hooks with big eyes, such as the Daiichi 1110, a favorite of mine for years. Look through your dry fly box and see if you have other styles of flies with materials that cover the hook eye. If so, tie a few of those patterns on big eye hooks, keeping them reserved in your box for situations where you have to thread the tippet through in a hurry or are having problems seeing the opening.

✔ Fishing Suggestion

As I travel around the country presenting to various trout clubs and organizations, one of my most popular topics is emergers. Many of my largest fish caught on top have fell victim to an emerger fly, and this one has been a go-to during Pennsylvania's famous Sulphur hatch. Once I have rising trout, especially those that appear "splashy," an emerger is the first topwater fly I choose, and it rarely comes off until the hatch concludes. Fish become programmed to eat easy protein requiring little effort, and an emerger guarantees that the fly is stuck there for a while. Think of this like a trout waiting at the end of a conveyor belt that is constantly popping peanut butter cups into its waiting mouth, it can be that easy for a trout during the emerger stage. Instead of a traditional parachute fly, choose one that represents this specific stage and find the time(s) it works best for you . . . then give those trout some peanut butter cups!

Tying the Condor Emerger

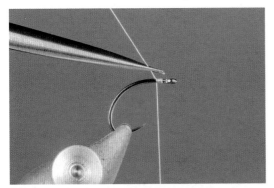

1. Place the hook in the vise, then start the thread approximately two wraps before the eye. Wrap rearward a few turns, then trim the thread tag. I prefer to have the straight section of hook parallel to the ground during most tying steps. Feel free to move the hook placement within the vise to one that suits your tying, especially for emerger patterns.

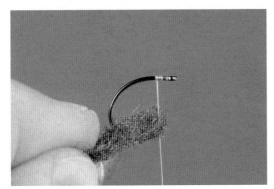

2. Cut a small section of Antron, then double it around the thread. Hold the Antron ends together with the pointer finger and thumb

of your left hand, then gently pull them tight toward yourself. This placement allows you to control the bobbin with your right hand for the next step.

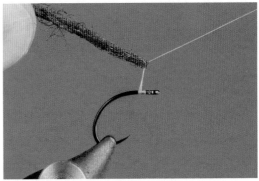

3. Wrap the thread around the hook while holding the Antron tight. As your thread makes it over the top of the shank, continue pulling the Antron gently to the left and away from the thread direction, helping to maintain tension.

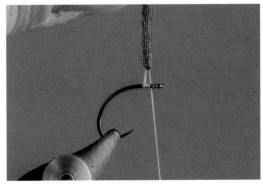

4. Complete the thread wrap around the hook, continuing to hold the Antron directly above the hook. As you pull down on the bobbin, allow the Antron to gently pull toward the top of the hook shank.

5. Once the Antron rests against the top of the hook shank, place two thread wraps in front of and behind the post, securing everything in place.

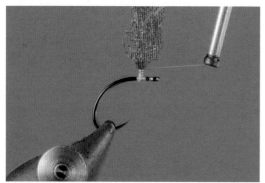

6. Begin to create a parachute post by making helicopter thread wraps around the Antron, working up in a clockwise direction (when viewed from the top). Be sure to make touching thread wraps as you wind. Notice how the bobbin tip is facing down and close to the post; doing so will make the technique easier to master.

7. Continue wrapping the thread until you've gone up the post approximately the same distance as the post is from the hook eye. This will provide durability and stability for the post and allow the hackle to lie smoothly against the Antron. Wrap the thread with touching wraps back down the post, then place three crisscrossed thread wraps around to finish securing. (There are many parachute tools on the fly-tying market to assist with this technique, though with time, it becomes much easier without a reliance on a tool.)

8. Cut a small pinch of fibers for the trailing shuck and secure by the post with three thread wraps. If the tag of the material is extending past the post, gently pull on the end by the hook bend, pulling the fibers close to the post. This will save you a step by not having to cut the tag ends of the shuck. *Tying tip:* If using straggly fibers, wet slightly with water to help them stay in place before securing with thread. This tip has rescued me many times when tying with a variety of both natural and synthetic materials.

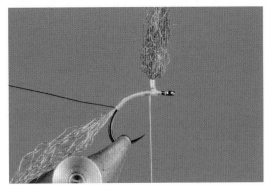

9. Continue the thread with touching wraps toward the rear of the hook, stopping as the shuck fibers begin to bend significantly downward. While wrapping, hold the trailing shuck ends in your fingers, keeping them bunched together on top of the hook shank. This is a match-the-hatch type of pattern; consider using a hi-vis trailing shuck as a hot spot for a different kind of appeal to the fish.

11. Wind the thread forward with touching wraps to the parachute post, ensuring that the thread is even from shuck to post.

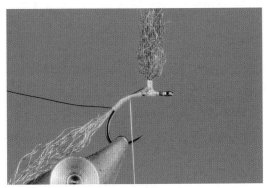

10. Snip a short piece of wire and lay it on the body of the hook. Allow the butt end of the wire to nearly touch the parachute post, then secure it at the shuck with a pinch wrap and multiple turns of thread. Begin winding the thread forward toward the parachute post.

12. Wind rearward with three touching wraps, then tie in two of the condor substitute fibers by the tips. These are fragile fibers, especially by the tips; be very intentional with your thread wraps.

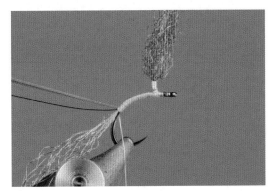

13. Lock the condor substitute fibers to the shank with touching wraps, wrapping the thread completely to the shuck. At this point, the condor substitute and wire should be at the same location.

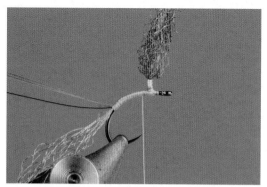

14. Wind the thread forward with touching wraps, returning to the original tie-in location of the condor substitute. If a slight taper is desired, this would be the step to create one. Build a taper that increases in diameter as it approaches the parachute post.

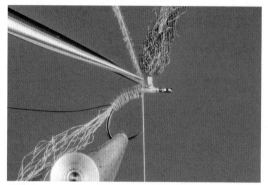

15. Wind the condor substitute forward to meet the thread, then secure with three thread wraps a few turns short of the parachute post. Trim the butt ends of the condor substitute.

16. Counter-rib the wire through the condor substitute with evenly spaced wraps. As you wind forward, gently wiggle the wire back and forth, which will encourage it to sit within the condor substitute, versus matting down the fibers. Secure the wire close to the parachute post with three thread wraps.

17. Place one thread wrap in front of the wire, between it and the post. Hold the bobbin down with slight pressure, then helicopter the wire away. To helicopter, lower the wire so it is nearly parallel to the fly's body, then move it in a circle. When done correctly, and holding the bobbin under tension, the wire will break away cleanly.

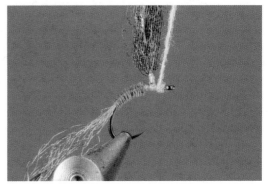

18. Prepare hackle by stripping barbs off of the butt section of the stem. The remaining fibers should all be approximately the same length. Next, tie in the stem against the base of the parachute post, with the cupped side of the hackle facing up.

20. Create a short dubbing noodle on the thread, then wrap rearward toward the abdomen. Maintain tension on the thread and use touching wraps to prevent any gaps in the thorax. **Note:** I prefer a tightly dubbed thorax for most dry flies, but when tying and fishing emergers, straggly fibers are acceptable and encouraged!

19. Gently pull the hackle up, with the feather lying parallel to and nearly touching the parachute post. Keeping the bobbin tip pointed down, begin tying up the post with touching thread wraps, locking the hackle against it. Be sure to keep the cupped side of the hackle facing the post. Return down the post with the thread, then advance the thread to nearly touch the hook eye.

21. Continue wrapping the dubbing noodle around the parachute post and to the rear of the thorax. Once you've filled the empty section of hook and examined all sides of the body, remove any excess dubbing from the thread. At this point, the thread should be hanging between the abdomen and the thorax.

22. Lift the bobbin backward, as if going to unwind a wrap, but before completely rotating the thread around the hook, make a thread wrap around the parachute post, using a helicopter method.

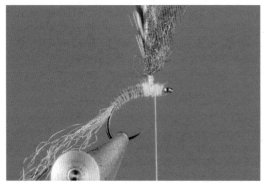

23. After one thread rotation around the parachute post, allow the thread to lie into the thorax on your side of the hook. This will be its resting point until needed to tie off the hackle.

24. Attach hackle pliers onto the tips of the hackle, then begin to wind with touching wraps down the post. Gently wind the thread up and down when wrapping, allowing it to sit within the fibers (versus matting them down). My preference is clockwise wraps with the concave side of the hackle facing up, which allows more of the fly's abdomen to rest in the film. *Tying tip:* Hackle placement is personal and varies based on how you want the fly to rest. There are times I want the hackle tips to poke through the water's surface and will reverse it to make the concave side point down.

25. Lift the bobbin on your side of the fly, unwinding the thread a half turn. Next, make a helicopter wrap with the thread around the parachute post, locking the hackle to the post (notice how the bobbin tip is face down). Continue making multiple thread wraps both above and below the hackle tip while holding it securely with the hackle pliers.

26. After five thread wraps, jump the thread from the hackle tie-off location at the post to the hook eye. The thread will be lying across the top of the thorax and invisible from underneath the hook (to both fish and fishers). Gently pull the parachute post to the left, exposing more of the hook eye. Whip finish at the eye, trying your best not to allow any hackle fibers to catch in the thread.

27. Trim the thread tag, then use your scissors to cut the hackle tip close to the parachute post. When deciding on a post length, longer is not always better, as it will encourage your tippet to spin when casting. On parachute patterns, I tend to cut the post so it's slightly larger than the hook gap, then trim on the water as necessary. This specific imitation was tied to represent Sulphur and Pale Morning Dun mayflies, though with slight color variations, the pattern can imitate many other mayfly and caddisfly emergers, so be sure to identify the local insects on the waterways you fish and match accordingly.

Pliva Shuttlecock

Shuttlecocks are a popular pattern in the UK and throughout Europe, as the CDC encourages their abdomen to sit within the surface film. When I find rising or sipping fish in slow-moving water, this is one of the first patterns that I add to my tippet.

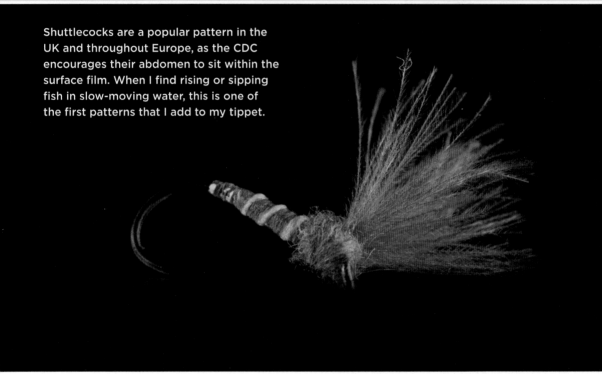

Tying and fishing delicate flies can be a chore, especially those incorporating CDC, but here's a pattern involving various techniques that you can apply to many others in your arsenal. There are a number of unique techniques and materials (even superglue) in this emerger, all used to enhance the body's durability and effectiveness. When combined with a tag hot spot, corded thread ribbing, and hi-vis CDC, this pattern encompasses a range of tying skills to master.

The design of this shuttlecock pattern encourages the body to sit under the surface film, while the CDC fibers keep the wing in view for us. This overall profile is intended to mimic the emerger stage, an insect transitioning from nymph to adult. Fish know that the insects are vulnerable during

Pliva Shuttlecock

- **HOOK:** #18 Hanak H 130 BL
- **THREAD:** Olive Semperfli Spyder 18/0
- **TAG:** Pearl Midge Flash
- **RIBBING:** Glo-Brite #12
- **ADHESIVE:** Fly Tyers Z-Ment
- **WING:** Three CDC fibers (two natural, one dyed)
- **THORAX:** Adams Gray Superfine Dubbing

the change and capitalize by feeding at the surface . . . which we love as fly fishers! The design of this fly is credited to Devin Olsen, a member of Fly Fishing Team USA. He has had incredible experience with fish all over the world, and his Pliva Shuttlecock design incorporates lots of the essential elements needed in an effective emerger pattern.

Let's start by breaking this pattern down into its more prominent features. When examining the finished picture, notice what first catches your eye; for me, it's the unique tag, which is fine tinsel. The tag in these situations can be a hot spot or represent a trailing shuck, and its main objective is to attract the fish, just like it grabs *our* attention. After wrapping the tinsel in place, be sure to carefully inspect it, especially keeping in mind that when you use transparent materials, the thread

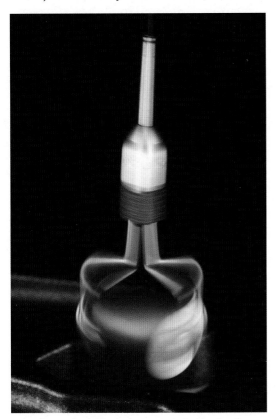

A look at a bobbin being spun clockwise to cord the tying thread. Knowing when and how to employ this simple technique can help you move from a beginner to an intermediate tier.

color underneath may lighten or darken the overall effect. You can use this to your advantage on other patterns, experimenting to determine the right color shine.

Next, be careful not to overlook the ribbing on this fly, a contrasting color of thread that has been corded. This simple technique takes only a few additional seconds, and the prominent ribbing that gets created truly gives the impression of body segments. I encourage you to experiment with ribbing colors to see what works best in your local waters. A good starting place is to try fluorescent colors that contrast with the body, especially when fishing off-colored water with faster current. For slower current and clearer water, choose colors that are closer together on the color wheel.

Superglue comes next, and years ago I would have never imagined its place of prominence on a dry fly, let alone one I'm sharing in a published book! As we continue to develop flies in the 21st century, I've noticed an increased emphasis on durability. There are many factors encouraging this, such as the competitive fly-fishing world and anglers who want to use a pattern that maximizes the amount of time spent fly fishing (versus replacing a fly that is falling apart). Dry flies are no exception, and the superglue soaks into the body here, giving the desired durability to the thread body, while also leaving a slight sheen that suits the emerger stage of this fly. UV resins have their place; however, they can build up quickly on the outside of the pattern, causing it to sink below the film into an undesired water column.

On to the CDC, which even as a delicate material can accumulate quickly on smaller flies. Notice that during the tying, an extra step is taken to remove the center stem of each CDC feather tip, ensuring that you will only have those perfect CDC fibers to aid in flotation for your pattern. Another advantage comes into play with smaller flies: After cutting the CDC butt section, place it in a material clip, because it can be used on your next fly . . . even when those fibers have been trimmed with scissors (read more in the CDC Tying Tip section).

Finally, hi-vis materials are nearly an automatic incorporation in my tying, both as hot spots and to aid in visibility. Leaning on the latter, the dyed

CDC contrasts nicely against its natural brethren, helping us see this fly at distance, especially in low-light situations. Think about some of the patterns you tie that sit lower in the water, and consider using a hi-vis material in them; selecting a natural material such as dyed CDC will help keep your flies lightweight and floating longer.

✔ Tying Tip

CDC is an incredible material that belongs on the tying desk for many types of patterns, from the obvious dry flies, through emergers, and even in nymphs (see the CDC Quill Body Jig Nymph pattern in this book). With its ability to impart movement, more of my fly designs incorporate using CDC for wings, legs, and as a body material. These quick tips will enhance your tying with CDC:

- Be careful with adhesives around these fibers and others, as their use can result in an unintended consequence, such as this one shared to me by Devin. When thinking about changing the order of materials, I considered tying the pattern through the CDC wing, and then applying superglue to lock everything in place prior to dubbing the thorax. Devon reminded me that superglue has the potential to soak into the base of the CDC fibers, making them hard and brittle; instead the adhesive is added to soak into the body before the wing is secured with thread.
- In tying, the hard edges created from scissors when trimming CDC barbs appear unnatural at times (more so to us than the fish), but by pulling the tips between your thumb and pointer finger, it's possible to tear the CDC fibers and make them appear more natural.
- There are many ways to incorporate CDC, including wrapped as a soft hackle, spinning the fibers in a dubbing loop, or post-style like this pattern. Experiment with this "magical" material, as it's one that will enhance your patterns . . . and bring you a few more fish!

✔ Materials to Consider

After persuading you to expand your collection of CDC, dial it back before buying all those dyed colors. First, check your tying collection for hi-vis materials that can be incorporated into the wing. Many parachute post materials are waterproof, helping to keep the wing above the water. They also come in a variety of fluorescent colors and are reasonably priced. The added bulk over a piece of dyed CDC is minimal, so this may be a great starting point for you. Unfortunately, if you're anything like me, you'll also want nearly *every* dyed CDC color known to mankind!

✔ Featured Technique

Spinning the bobbin in various directions will either bind your thread together, commonly referred to as cording, or the opposite, uncord and flatten it. Illustrated on this emerger, cording is a simple way to create a prominent rib that I encourage you to try on your dry flies and nymphs, too. Other advantages to cording thread include an increase in its strength and the ability to encourage the thread to "jump" one way or another when winding. Be careful . . . not all threads are created equal! Don't spin the bobbin or tool too much or else the thread will break, a disappointment for all of us when in the middle of tying a fly. This basic technique will help you jump from one tying level to another; play around with it a bit and experiment how your threads (and other materials) work when cording.

✔ Fishing Suggestion

Emerger patterns are used when fish appear to be eating flies closer to the surface of the water. This one sits in the surface film and represents an insect transitioning from the nymphal to the adult stage. Appearing almost "stuck" between the two stages makes the emerger vulnerable; as a result, fish tend to key on it during hatches. This pattern represents the Blue-Winged Olive mayfly, but you can vary the body color based on other natural insects in the waterways you're fishing.

When fishing with emerger patterns, I prefer a drag-free drift. Cast the fly as close to the rising fish as possible without spooking it, guaranteeing a presentation with little drag. Other methods to reduce drag include:

- Adding additional or finer tippet
- Making a reach cast
- Placing yourself in a location that requires less fly line on the water

Imparting a slight twitch may give the impression of life, but subtlety is key! As your emerger begins to sink nearly out of sight, make a couple of additional drifts before applying floatant. Fly fishing with emergers is technical, but once the code is cracked, the end result is well worth the time and effort.

Tying the Pliva Shuttlecock

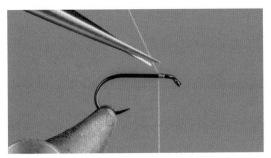

1. Place the hook in the vise, with the shank parallel to the ground. Start the thread a few wraps behind the eye and make three touching wraps rearward, then trim the thread tag.

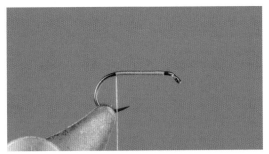

2. Continue making touching wraps rearward, completely covering the hook shank. Stop wrapping before the thread reaches the hook bend.

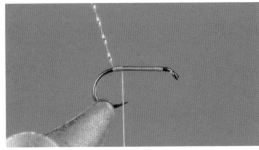

3. Tie in the Midge Flash at the bend with three wraps, then make four touching wraps forward. Trim any remaining flash material. *Tying tip:* Many times, it's recommended to lock a material in with loose wraps, then pull on the standing end until the tag ends are under the thread wraps. With certain materials, such as Midge Flash, doing so will stretch the material and change its overall appearance. Experiment to determine how much tension each material can handle.

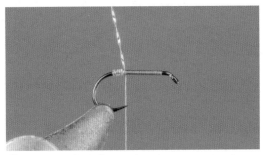

4. Advance flash forward with approximately four touching wraps. Secure with thread, then trim the tag ends of the flash.

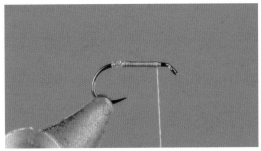

5. After securing the flash, advance the thread forward, stopping just short of the hook eye. Ensure that the body is relatively smooth at this point.

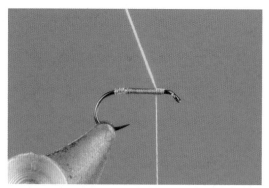

6. Lock in a single strand of Glo-Brite with the tag end facing the hook eye. Trim the tag end (or pull gently under the thread wraps). Wrap the thread toward the rear of the hook with each turn.

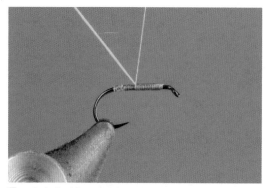

7. While holding Glo-Brite taut with your left hand, make touching thread wraps toward the rear of the hook and lock the Glo-Brite against the top of the hook shank. Keep the fly's body uniform and slender while wrapping.

8. Lock the Glo-Brite in place against the Midge Flash tag, and try not to cover the flash with any thread wraps.

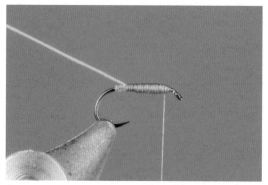

9. Create a slight carrot taper, wrapping the thread to the hook eye and returning to the abdomen and thorax with successive wraps. Finish wrapping the thread and stop close to the initial tie-in location. (*Note:* A carrot taper consists of a body that is narrower by the tail and wider toward the bead.)

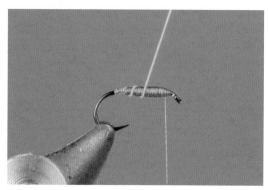

10. Cord the Glo-Brite by spinning it clockwise, either with hackle pliers or by hand. Once tight, begin wrapping it forward, ensuring even spacing. In the picture, you can see how the ribbing adds a dimension (versus lying flat). This comes from cording the material and maintaining pressure while wrapping.

11. Continue wrapping the ribbing to the waiting thread, then secure it in place with multiple thread wraps. Be sure to keep consistent pressure on both the Glo-Brite and the thread during this process.

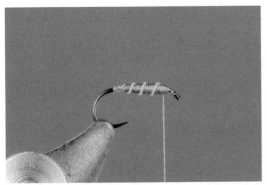

12. Once you've ensured the ribbing is evenly spaced and secure, trim the tag end.

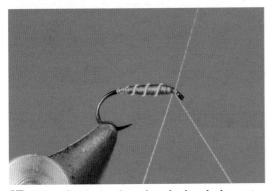

13. Whip finish the thread at the head, then trim the tag end. You can leave the thread in place, but be sure to work around it in the next few steps or else the standing thread will be nearly brittle and difficult to wind in future steps.

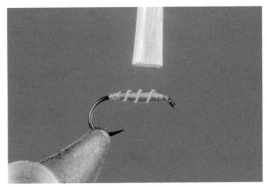

14. Using superglue, dip the brush into the bottle, then allow the excess to drain back in. As I remove the brush from the bottle, I will also dab the fibers against the inside neck, so there is not a blot of superglue at the end of the brush.

15. Apply a thin coat of superglue to the thorax of the fly, taking time to completely cover all sides. I prefer to work from the eye rearward.

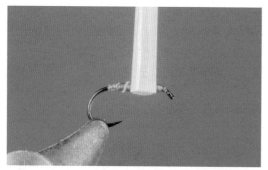

16. Continue the application of superglue into the abdomen and toward the tag. Ensure that every section is covered with a thin amount of adhesive. If you notice that the superglue on the brush is excessive, dab it on a piece of paper or Post-it Note on your tying desk.

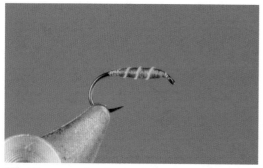

17. Allow time for the superglue to work . . . and try to resist touching it to check! While waiting, I recommend working on more of the same style of pattern, tying a handful to this point before continuing with the final steps.

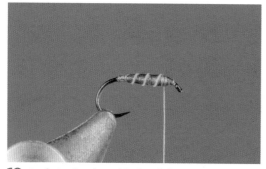

18. Lash in the thread behind the eye with a few wraps, then trim the tag. Prepare pieces of CDC by cutting away the center stems near the tip of each feather.

19. Tie in the CDC with the tips extending a hook length past the eye. All the CDC fibers should be on the top of the hook shank. *Tying tip:* If using a hi-vis CDC feather as a sighter, I recommend stacking that color on top during the initial tie-in. Doing so will allow it to be seen by you, but not in the "front and center" position potentially visible to a rising trout.

20. If the tips extend too far, you can pull slightly on the butt ends until reaching the desired length. If doing so, ensure that the CDC is lashed in tightly with the thread afterward. In the pattern tied for this book, you'll note three CDC feathers total, two natural and one hi-vis.

21. Hold the butt end of the CDC fibers perpendicular to the hook shank, allowing access to the tie-in location for the scissors.

22. Make a neat cut with your scissors, removing the butt ends of the CDC close to the fly's body. I prefer to cut with the fine scissor tips, though I know of many tiers (with very steady hands!) that choose razor blades for this, too.

23. Take a pinch of dubbing and noodle it between your finger and thumb onto the thread, spinning the dubbing in one direction. Use fine dubbing for this step to create a tight and sparse noodle.

24. Slide the dubbing up the thread so it's nearly touching the fly, then begin to wrap the dubbing noodle over the CDC butt ends, completely covering them.

25. Continue the dubbing noodle forward with touching wraps toward the hook's eye. Stop when reaching the CDC fibers, as you don't want to mat them down, but just cover the thread wraps.

26. Pull the CDC fibers perpendicular to the hook shank and place wraps with the dubbing noodle directly at their base. This dubbing dam should be tight against the CDC, helping the fibers maintain an upright position. Continue wrapping the dubbing forward toward the eye.

27. Remove any dubbing from the thread, then whip finish at the eye. Trim the tag end of the thread tight against the hook.

28. If the CDC tips are uneven (and that bothers you), grab on to their base with the pointer finger and thumb of one hand, and with the other, slowly pull the tips to tear off. Many tiers, including me, use scissors for a fast result . . . though this may seem sacrilegious. Be sure to vary the body color on this pattern based on the naturals in your waters and try different materials and colors for the hot spot tag. Adding a short Zelon trailing shuck is another great choice to mimic vulnerability in the emerger stage.

NYMPHS

Frenchie

A hybrid fly that grew out of the Pheasant Tail's fame, the Frenchie style has inspired countless variations that have fooled trout for me in both moving water and stillwater situations.

A commonly asked question in fly fishing and tying goes something like this: "Which nymphs should I start with when getting into fly tying?" The reply is nearly automatic, beginning with a Pheasant Tail or Hare's Ear (sounds familiar, right?). Both have qualities that can represent many mayflies and caddisflies, but let's examine these from a modern-day approach.

Starting with the former, commonly referred to as a PT, the Frenchie is the modern-day equivalent to Frank Sawyer's pattern, and it's a fly that has morphed since its inception. Learn the basics of the style, and you'll be able to apply the technique to many nymphs currently being tied, especially those on jig hooks.

Frenchie

- **HOOK:** #14 Hanak H 450 BL
- **THREAD:** Fire orange UNI 8/0
- **BEAD:** Gold 2.5 mm slotted tungsten
- **TAIL:** Medium pardo CDL
- **RIBBING:** Gold UNI-French XS
- **BODY:** Pheasant tail fibers
- **THORAX:** Bighorn orange Sow-Scud Dubbing

While fly fishing in Wyoming with Rob Giannino, I expected some dry-fly action from the start. The water hadn't warmed yet, so we fished the nymphal stage of the Blue-Winged Olive life cycle and were rewarded with some gorgeous trout. COURTESY OF BLACK MOUNTAIN CINEMA

A significant change from the original PT is the tail, with this version using Coq de Leon, a resilient fiber known for its sheen, mottling, and durability. The latter factor is a key reason why it replaces pheasant tail fibers, which tend to tear easily, especially after catching many trout. Another added benefit is that fluorescent dyed versions of CDL are becoming more readily available, which makes a sneaky choice for a rear hot spot. The body maintains the traditional component, as the wet PT fibers appear like gills in the water. But wait, didn't I just talk about their fragility? Absolutely, which is why this pattern has a counter-wrapped ribbing material, in this case a soft and flexible wire.

The major component that sets this pattern apart is a thorax tied with fluorescent dubbing that typically has some flash, seen as a hot spot. This is an attractor pattern, and popular colors include hot orange, pink, and red. When selecting a color, I prefer to use one that creates contrast with the body, though there are times when I'll choose a color so obnoxious, the fish have to either commit or frantically swim away!

Not all dubbing is created equal, and those that I recommend for the thorax are typically synthetic, which can be a little difficult to dub. Remember when creating the dubbing noodle that you must spin the dubbing around your thread in the same direction, creating a thin noodle that will cover the thorax in a few turns. By "few" I honestly mean around three, and one to grow on, with more being overkill on this style of fly.

Finally, bead choice is something to keep in mind, and my thoughts lead down two pathways:

- Flashy. This is my go-to color choice, and silver and gold are my two favorite colors on the Frenchie. Depending on which I choose, when ribbing with wire, I tend to match the color.
- Subtle. When fishing in areas that put pressure on fish, this is the route I take. Colors such as matte black and even brass are perfect because they provide weight and truly create contrasting colors with the thorax sandwiched between two darker sections.

When using a fluorescent-colored thread, allow a few turns to show between the CDL tailing fibers and PT body, creating a tag hot spot. This difference may be enough to separate your fly from the others being used on the water today. When doing this, consider sealing the exposed thread with head cement or UV resin, as it can tear easily from the fish chewing the rear apart. Match this color to the thorax unless you're adventurous and select a second hot spot color. What would Frank Sawyer think about that?

✔ **Featured Technique**

After locking the tailing fibers in place, be sure to gently pull them toward you as the thread makes each turn. The thread will carry them over the hook, encouraging the fibers to lie flat and straight down the shank. Few things look better than CDL exiting the nymph at 180 degrees!

✔ **Materials to Consider**

A common and obvious modification to this pattern is varying the thorax colors, but also think about the different dyed pheasant tails that are now available. These choices can either more accurately imitate natural insects or create additional contrast with your pattern. Favorite colors include olive, red, orange, black, and even bleached. As I'm sure you've already guessed, yes, they even dye pheasant tails in fluorescent colors . . . and they are excellent choices when fishing for many species, especially steelhead.

✔ **Fishing Suggestion**

The Frenchie is a suggestive pattern with a body typically intended to imitate a mayfly nymph. The addition of the thorax hot spot makes it a great searching pattern for opportunistic fish, and it is a perfect starting fly when trying out the European nymphing fly-fishing technique. I prefer to fish this pattern in riffled water or at the heads of pools,

and there are many excellent fly fishers that carry *only* this fly in varying sizes, colors, and weights.

Tying the Frenchie

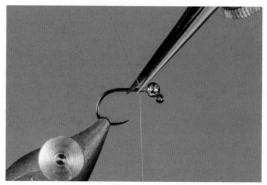

1. Place the slotted bead onto the hook, inserting the point through the smaller hole. Slide the bead into position against the eye, with the slot facing up. Begin with thread wraps jammed against the inside of the bead, building a thread dam tight against the bead so it won't move rearward. Make six touching wraps toward the rear, then trim the tag end of the thread.

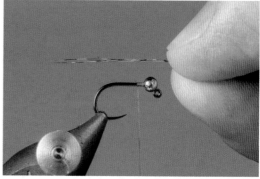

2. Measure the Coq de Leon (CDL) tips to be approximately the same length as the straight part of the hook shank, then place them over the hook with tips facing rearward.

3. Secure the tips with three thread wraps, allowing the CDL butt ends to extend over the bead. After securing, verify their length; if they are too short or long, lightly pull on the applicable end to correct.

4. Trim the butt ends, then begin wrapping touching wraps rearward with the thread. While winding the thread with your right hand, gently pull the tips of the CDL fibers toward you as the wraps progress. With each wrap, allow the thread to secure the fibers on top of the hook shank.

5. Cut a section of fine wire 3 inches in length, then place one end in the slotted bead opening. Allow the wire to rest on the thread body, then

secure it with three wraps against the shank. The location should be approximately above the barb of the hook. (Note that all hooks in this book are barbless, so that location is about one-third from the hook point.)

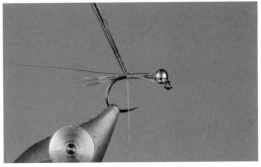

6. Remove about four pheasant tail fibers from the tail, then lay the tips so they are flush with the hook shank and extending into the thorax (nearly touching the bead). Secure with three wraps of thread at the "barb" location.

7. Wrap the thread to the beginning of the thorax, then grab the butt ends of the pheasant tails with hackle pliers. Wind the pheasant tail fibers forward to meet the thread in the thorax, then secure the fibers with five wraps of thread. Trim the butt ends flush against the body. *Tying tip:* Ensure that the first pheasant tail wrap covers the thread by making a full 360-degree wrap, then begin winding the fibers forward. On many patterns, it's easy to make that first turn forward without wrapping completely around, so be intentional when winding.

8. Counter-wrap the wire as ribbing to the thorax, going in the opposite direction of the pheasant tail fibers. Secure the wire with three thread wraps, then helicopter the fine wire away. To "helicopter," maintain firm pressure downward on the thread with one hand, and grasp the wire end with the other. Wind the wire like a helicopter blade in a clockwise direction.

9. Keep the thread at the end of the pheasant tail fibers, which starts the thorax of the fly. Take a pinch of dubbing and noodle it between your finger and thumb onto the thread. Distribute the dubbing onto the thread with the thin end of the noodle nearly touching the thorax. Begin winding the dubbing noodle toward the eye.

10. Finish wrapping the noodle directly behind the bead, with no more than three total wraps

of dubbing. If any dubbing remains on the thread, lightly pull it away and off so only bare thread remains.

11. Use a whip finish to secure the thread between the dubbing and the bead, making approximately five turns. For a more secure finish using superglue or head cement, brush a small amount directly to the thread prior to the whip finish. Trim the thread close to the hook.

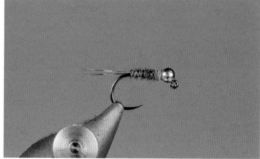

12. To vary, use dyed pheasant tail (I *love* olive) and change the dubbing colors, too. Favorites include fluorescent shades of orange, pink, and red. Many fly fishers dedicate an entire fly box to Frenchies, as their effectiveness cannot be overstated. Find the color combinations that work best for you and make Frank Sawyer proud!

CDC Quill Body Jig Nymph

CDC on a nymph? Yes! The movement gained from the addition of CDC is perfect for many styles of patterns, especially nymphs and wet flies.

Quill body flies have long been a popular choice for tiers and fly anglers, as their natural ribbing makes the pattern appear real enough to crawl or fly away. The addition of cul de canard (CDC), a delicate duck feather, gives this fly more lifelike characteristics, as those fibers undulate due to underwater currents. These two materials are typically considered essential for many dry flies, but we're going to integrate them into an effective nymph pattern . . . which you can apply to many more at the bench.

An initial problem for many is wrapping and securing the quill body, due to its fragile nature. A common tip is to soak the quills in water for a few hours before tying, making them pliable and easier to wrap forward. Many prepackaged quills

CDC Quill Body Jig Nymph

- **HOOK:** #14 Hanak H 450 BL
- **BEAD:** Silver 2.5 mm slotted tungsten
- **THREADS:** Brown and fire orange UNI 8/0
- **TAIL:** Medium pardo CDL
- **BODY:** Polish quill
- **UV RESIN:** Solarez Bone Dry
- **THORAX:** SLF Brown Spikey Squirrel
- **WING:** Natural CDC

in today's market don't have that need, as you're paying for someone to preselect the best for you, ready to go. After tying them in, I prefer to use hackle pliers with soft tips, as there is nothing more frustrating than wrapping one forward and having it break right before the locking wrap! Once the quill is locked in place, I turn to Solarez Bone Dry UV resin, which has a short set time, less than 15 seconds. Counter-ribbing is another option, though I prefer the glossy appearance given by resin, allowing the quill body's pattern to become a focal point and capture attention (hopefully of a trout!).

Returning to CDC feathers, get used to seeing them featured more regularly on nymphs. Many tiers value the feathers for the flotation they offer dry flies and the gentle movements given to emerger patterns. We are capitalizing on the latter, as the fibers are extremely delicate and move with each underwater current they meet. Using them in place of traditional soft hackle feathers (think Hungarian partridge and grouse) adds a new dimension to fly tying. As with many replacement materials, there are pros and cons, and in this one, we gain a lifelike ability but lose mottled fibers that replicate insect legs. In my opinion this is a positive trade-off, yet I continue to tie with both types of feathers in soft hackle situations.

Many options for peacock quills exist, both naturals and synthetics. An added advantage to us as tiers is that the natural quills are also dyed to provide a range of options for the body of nymph and dry-fly patterns.

Breaking this pattern apart, we have returned to Coq de Leon, again for its durability and mottled appearance, with additional tailing options including wood duck, mallard flank, and pheasant tail fibers. Feel like skipping a step? If so, lose the tail fibers and leave the CDC extending past the bend of the hook, which gives the pattern the appearance of a caddis.

Quill body patterns simply have that "look" to them, and don't be deterred if it takes a few (dozen!) until they begin meeting your expectations. Aside from the tips shared above, consider using a form of magnification for wrapping the quill forward. Each wrap should slightly cover the previous, and a bright light and cheaters help ensure this process goes smoothly. If you're having trouble mastering the body, there are synthetic quill substitutes on the market to help.

An important, yet often overlooked, step is adding a turn or two of dubbing before the CDC, which helps keep the material from folding against the body when wet. Experiment with the type of dubbing for this step, as you can go the subtle route, like the brown used here, or opt for one with more sparkle, such as an Ice Dub.

When using CDC as a hackle, you'll notice I chose one with a stem running up the center. There are other types of CDC available on the market, each with its own nuance and possibly requiring differing methods to secure and wrap. When selecting those without a firm stem, a common technique to consider is a dubbing loop with the CDC fibers. Additional steps require additional time and sometimes different tools, which is why I prefer those feathers with the stem (and spend that extra time fishing!).

✓ Tying Tip

UV resins have recently taken the fly-tying world by storm, and the use of Bone Dry here is two-fold. The resin hardens quickly, sealing the quill in an almost bulletproof casing. Once cured, the markings on the quill really jump out, and we can further accentuate them by overlaying brightly colored ribbing or tinsel prior to the resin application. The trickiest part of using resins is that a little goes a long way, and in most cases less is more. When applying Bone Dry and other resins, I either use the applicator brush or my bodkin, then rotate my vise before curing. If I notice an excess amount, a small piece of paper comes to the rescue, and I'll use it to dab away small amounts of resin.

✓ Featured Technique

Tying CDC by the tips is an intermediate technique that can be mastered in a short time by ensuring that you properly prepare the feather. Stroke the fibers down and away from the stem, leaving a small bundle to use as the tie-in point. After securing the stem against the shank, snip away the tip then palmer the fibers as you would Woolly Bugger hackle . . . but only make a couple (*two!*) turns; the quill builds up quickly. More wraps are welcome on dry flies when flotation is a concern, but for nymphs we want a handful of CDC barbs to dance with the water currents. Finally, if you use superglue to typically seal your thread wraps, opt for something else with CDC, as it can seep into the fibers, hardening them quickly.

✓ Materials to Consider

There are many methods to stripping barbules from peacock eyes, some more tedious than others. The preferred route is to purchase Polish quills, which are stripped and come in a variety of colors. Synthetic quills are becoming more realistic, offering a durable solution if their natural counterparts are too delicate to work with. Some even offer a clear quill, allowing you to vary the thread or tinsel color underneath, making for some fantastic color combinations with this pattern.

✓ Fishing Suggestion

With a slotted tungsten bead, you'll be fishing patterns like this near the bottom of the water column in rivers. Remember that the hot spot is not required, and I carry both versions in my box. Sans the hot spot, this is one I recommend using once you've identified fish in holding spots, specifically those in slower-moving water. That additional time

allows the CDC fibers to undulate, movement that gives fish the impression of life. When tying, be intentional in your selection of materials, thinking ahead to possible situations in which they have the potential to outperform other patterns.

Tying the CDC Quill Body Jig Nymph

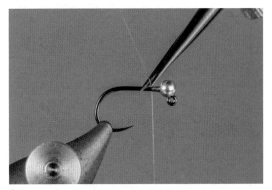

1. Place the slotted bead onto the hook, inserting the point through the smaller hole. Slide the bead into position against the eye, with the slot facing up. Begin with thread wraps jammed against the inside of the bead, building a thread dam tight against the bead so it won't move rearward. Make six touching wraps toward the rear, then trim the tag end of the thread.

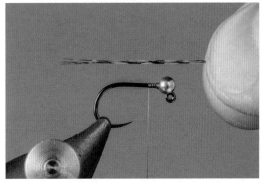

2. Trim or pull four to six fibers of Coq de Leon from a feather and measure them against the hook shank. The tips should extend past the bend of the hook approximately the same length as the body.

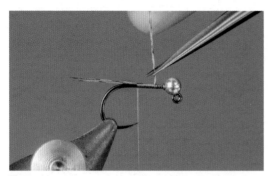

3. After measuring, transfer the fibers to your left hand, keeping the tips pointing away from the hook rear. Secure the fibers with three pinch wraps, tight against the hook shank. Trim the butt ends, then begin wrapping toward the rear of the hook.

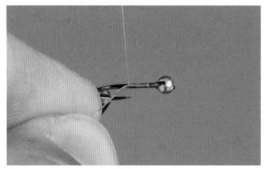

4. While winding thread with your right hand, gently pull the tips of the CDL fibers toward you as the wraps progress. With each wrap, allow the thread to secure the fibers on top of the hook shank. Stop wrapping before the hook begins to bend down toward the gap.

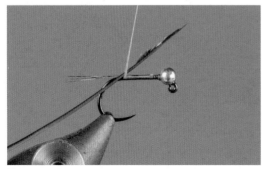

5. Select a Polish quill and stroke any remaining fibers toward the butt to remove; those by the tip may remain. Tie the quill in place by the tip with

two wraps, with its butt going to the left. It is my preference to have the dark section of the quill on the left side when securing, which gives the finished look you see in the abdomen.

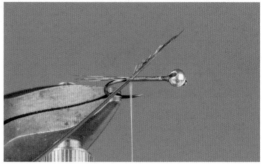

6. Holding the butt end of the quill, slide it toward you until the tip fibers get to where the thread was secured. This will maximize the amount of usable area for the quill. Be aware that the quill tip can be fragile, so I prefer to leave some remaining, which also helps build the body taper.

7. Advance the thread forward, building a slight taper that is narrower by the tail and wider toward the bead. Trim the remaining end of the quill close to the bead, then cover it with thread wraps. To build a full taper, I prefer to fully wrap the thread to the bead, then return approximately three-fourths of the way to the rear. Then I wrap to the thorax and back halfway to the rear. This back-and-forth wrapping will allow the taper to progress slightly.

8. Grip the butt of the quill with hackle pliers, then begin wrapping forward. Each wrap should slightly cover the front of the previous one, which encourages a segmented look from the quill body and dark line. Continue wrapping into the thorax.

9. Holding the pliers straight up, make a thread wrap over the quill. Attempt to angle the thread wrap toward the rear, which will help securely lock the quill in place. A common mistake is when tiers angle the quill forward before tying off, which can cause a gap or the last wrap to appear crooked compared to the other quills. Make two more wraps to secure the quill.

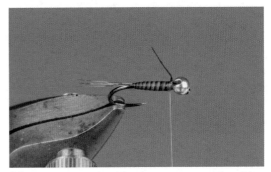

10. Remove the hackle pliers, ensuring that the quill segments are spaced to show segmentation and are angling in the same direction. If not, secure the quill butt with hackle pliers and unwind the thread, then rewrap the quill as needed.

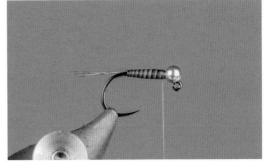

11. Once the quill is uniform, spaced evenly, and secured with thread wraps, cut the remaining quill flush to the body.

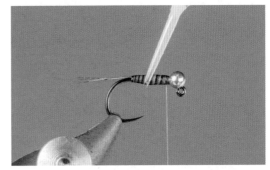

12. Apply UV resin evenly over the entire body, staying away from the tail fibers. Once the body appears completely coated, pause and verify that there is no excess drooping below the body. If so, lightly dab with a Post-it Note to absorb the excess resin.

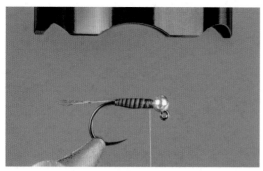

13. Cure the resin with a UV light, shining it on all sides of the fly based on manufacturer's recommendations. *Tying tip:* When curing UV resin, I place my hand between the fly and me, which prevents UV rays from reflecting into my eyes.

14. Take a pinch of dubbing and noodle it between your finger and thumb onto the thread. The noodle should completely surround the thread, with its end nearly touching the thorax. Begin winding the dubbing noodle toward the eye.

15. Wind the dubbing forward with touching wraps so that the thorax is uniform. This should be approximately five turns, stopping one turn

before the bead. Remove any excess dubbing from the thread.

16. Prepare the CDC feather by stroking the main fibers toward the butt, leaving the tips facing forward. Place the butt end of the CDC in your hackle pliers, as it will be tied in by the tips.

17. Gently apply water to the tips of the CDC feather, then run the fibers through your fingers. The tips will mat together, allowing your thread to maneuver without catching them when locking in.

18. Secure the CDC feather by its tip, locking it in place with three wraps. Ensure that the tips are facing forward, then trim them away.

19. Wrap the feathers around the hook two times. With each turn, fold or gently pull the fibers on both sides of the stem to the left side. This will encourage them to flow naturally to the rear of the hook, similar to the legs of the natural insect.

20. After two turns of the CDC, gently pull the fibers already wrapped rearward. Next, tilt the hackle pliers over the bead, which will expose a point on the stem to secure with multiple turns of thread. Once done, remove the hackle pliers, then trim the CDC butt close to the hook.

21. Again, pull all remaining fibers rearward, then take two turns between them and the bead. If going for a more natural look, at this

point apply a pinch of the same dubbing to the thread, noodle, then cover the thread wraps with dubbing. Whip finish and trim the thread.

22. If any CDC fibers are errant, pull them toward the rear, then lock in place with thread. Whip finish and trim the thread close to the hook.

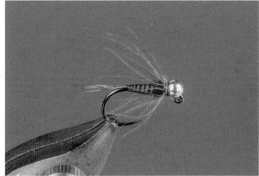

23. Inspect the CDC fibers, removing any errant ones. For those longer than the tail, shorten by either cutting or gently tearing. To tear, grasp the barb from the right side with your right thumb and pointer finger, then grasp its tip with your left thumb and pointer. Bring all fingers together to the point of the desired tear, then pull in opposite directions, tearing neatly between your fingers.

24. Tie the fluorescent thread in with a few wraps, then trim the tag. Continue wrapping until the space between the CDC and the bead is filled, then build up to the approximate level of the dubbing. *Tying tip:* If the round part of the top of the bead is flush with the thread, you've applied too much. Placing that much allows the hot spot to be easily torn when fishing.

25. Apply a thin amount of head cement directly to the thread, then whip finish and trim the tag. Afterward, an option is to apply a thin layer of UV resin around the hot spot to strengthen it, though be careful not to get any in the CDC fibers. My preference is to brush cement onto the thread, which will strengthen the overall hot spot and not impact the CDC.

26. Inspecting the finished fly, I find a lot to love here. Applying CDC to the front of many nymphs will give increased movement and the appearance of life. The added hot spot helps grab the attention of fish, and can vary based on colors that perform well in the waterways you fish. When varying, the first change I make is to select different colors for the abdomen, especially fluorescent colors. Use this pattern as a base to build from, and you will easily fill a fly box full of variations that will fool trout around the world.

Mop Fly

The infamous Mop fly! This version is tied with a natural body color and dubbing collar, but the real magic is what occurs as you are locking everything into place.

Wow, do I have to tread lightly with this fly . . . it is a fly, right? During the early stages of writing this book, Mr. Jay Nichols, the incredible publisher and brain behind it, was insistent that I locate and share flies from which others can learn techniques. Not intended to be a pattern book, the purpose of *Fly Tying for Everyone* is to enrich the reader's tying, teaching new skills to help you get to your personal next level of fly tying. Then there's the Mop fly!

If you take any time to examine my YouTube channel over the years, you'll see various styles of patterns, from trout flies to those used in salt water. Interests ranging, I've tied everything from traditional Catskill dry flies to tube flies, fascinated by each and every style. Then I moved on to jig

Mop Fly

- **HOOK:** #12 Hanak H 400 BL
- **BEAD:** Copper 3.0 mm slotted tungsten
- **THREAD:** Black Semperfli Nano Silk 12/0
- **BODY:** Tan Mop material
- **THORAX:** Peacock bronze Siman Peacock Dubbing Fine
- **ADHESIVE:** Fly Tyers Z-Ment

While fly fishing in Iceland, I was told that Arctic char never met a dry fly they didn't like. Well, I quickly proved that wrong, but I did determine that they *loved* their Mop flies!

COURTESY OF BLACK MOUNTAIN CINEMA

nymphs, flies popularized by the highly effective world of Euro nymphing. It was during this craze (which I'm still currently in, among other styles!), the Mop fly was introduced to me.

When the time came for me to share an early version of the Mop on YouTube, I was honestly unsure if I wanted to do so. Releasing a video of this infamous fly had the potential to alter the dynamics and audience of my channel; thus I held it back for some time before eventually sharing

the video publicly. Since that day, the pattern has been embraced by my viewers, and it holds a spot in the top five most viewed videos I've ever created. This book shares a more recent version that I tie, intended to show ways to increase durability for patterns.

Are there techniques used in this fly? Absolutely, though depending on whom you talk to, there are about 1,000 different ways to tie it. Advanced techniques add additional materials, with my version incorporating a dubbing loop to allow a veil of fibers. In its simplest form, some Mop material is lashed to the hook, with a whip finish at the head. If that truly was how I tied them, this fly would not have a place here, so let's tease out some techniques used with the Mop.

Examining Mop material, you'll find chenille bound together with a thread core. The exposed strands are like soft fingers, and the piece offers a decent amount of cushion when squeezed. Originally, when lashing the material onto the hook, I preferred exposing the inner core, then feeding the Mop onto the hook. When it's eventually slid down the shank, the inner cords grip the thread-coated hook, sealing tight together at the head following the application of superglue. With newer procedures, the Mop is set onto the shank, then the body is wrapped with spaced thread wraps. Superglue is used to initially coat the shank, as this extra step provides excellent durability and extends the fly's longevity—my reason for the switch.

With some materials, such as CDC, applying superglue will cause them to overharden and become brittle, but in this case, we're strengthening a bond. Securing everything with thread helps keep the Mop in place and provides an additional tier of toughness. As we've discussed in this book numerous times, durability of patterns is a key characteristic; in this case, the fish really do eat these flies and we need the pattern to be resilient. Think about other flies that tend to fall apart with little use: Is it possible to apply an adhesive or additional material to create a more durable pattern?

✓ Tying Tip

Becoming a proficient tier includes mastering thread control; knowing when and how to apply pressure can mean the difference between a secure material and thread snapping! Keeping the Mop snug against the superglue and hook shank increases the fly's longevity, and here's a simple tip that I learned, which adds virtually no additional time.

In the fall of 2019, I spent time tying at the International Fly Tying Symposium with Mike Komara, former captain of the Youth USA Fly Fishing Team. After placing the Mop on the hook shank, Mike forms a thread dubbing loop, winding it through the Mop body. Keeping his fingers inside the dubbing loop allows Mike the ability to provide pressure when winding, guaranteeing the Mop stays snug against the hook and the thread stays buried within the fibers. The key to this working is that your fingers are sensitive and allow you to feel the applied pressure, which can be difficult to sense when wrapping with a bobbin. Mike secures the loop with his bobbin behind the bead, then whip finishes sans dubbing. I forgot about this one when I was typing earlier—make that *1,001* different ways to tie the Mop.

✓ Featured Technique

Adhesives play a role in pattern durability, and four used regularly are:

- UV resin: This adhesive is sealed *quickly* with a UV light, and has secured a place on my bench as a "go-to" material. Starting with patterns such as the Perdigon, I quickly realized many possibilities with this, such as creating layers of an abdomen or even sealing dumbbell eyes and parachute posts. Be careful, as the resin can build up with too many layers. Also, UV rays are required to set the resin, so if you are sealing something inside a pattern, there are other adhesives for the job.
- Superglue: Sealing faster than epoxy, superglue's role is when I want to ensure little buildup of adhesive. It tends to be absorbed into materials, which can provide lots of durability. A

downside is that certain materials, especially natural fibers like CDC, become brittle upon application.
- Head cement: Upon completing many patterns, an easy way to seal during the whip finish is by applying head cement directly onto the thread with a brush. Some bottles have one built in, and my current favorite product for this is Sally Hansen's "Mega Shine." With a single application, the thread is sealed tightly. After completing the head on various wet flies and streamers, multiple applications will provide increased durability and give a high-sheen appearance.
- Epoxy: Requiring more set time than newer adhesives, many tiers stopped using epoxy once UV resins became popular. But when sealing various materials, especially opaque ones such as stick-on eyes, epoxy is perfect for the application, as UV rays only penetrate when dealing with transparent materials.

✓ Materials to Consider

After tying and fishing with the Mop for a few years, I felt that a spot existed for a micro version, and eventually stumbled on it. While I played around with a variety of materials, a suggestion was offered when I was spending some time in Holsinger's Fly Shop, located in Central Pennsylvania: Try Ultra Chenille. Tying Micro Mop flies using the techniques above proved deadly, especially on pressured fish in low-water situations. As with many patterns I tie and fish, variations appeal to me for different reasons, but especially to experiment with something different and review the results. Not all variations end up as successes—in fact very few do. However, when everything comes together, you have a winner, and the Micro Mop happens to be one of those.

✓ Fishing Suggestion

Honestly, I had reservations with the pattern, and still do, as my early fly-fishing experiences were traditional, involving chasing prolific hatches while carrying Catskill-style dry flies in my forest-green Wheatley fly box (a Christmas gift from my parents—what more could a teenager want?!). Egg flies were only used during steelhead season, and the San Juan worm was best put to work during high-water conditions. If this is you, don't worry, I get it, I really do.

Fishing the Mop fly is tough to do wrong, but keep in mind that though it's a heavier pattern, the material causes it to sink slower than anticipated for a fly of this size. Oversizing the bead helps, and I know fly fishers sizing up to a 4.5 mm tungsten one, which is significant weight. This will mean that you have greater contact with the pattern, giving you a tendency to feel the take more often. Popular fishing methods including dead-drifting, swinging, and even slowly stripping the pattern, similar to a streamer. When the Mop is drifting in the current, irregular movements occur, which help to drive the fish crazy; encourage this by building slack into your leader system. In stillwater situations, I've let the Mop sink slowly to the bottom, then jigged it slightly; largemouth bass love this! Determine which color combinations work best for you and have fun. The fish *will* respond; trout, steelhead, Arctic char, bass, and even crappie have fallen victim to the Mop.

My intention isn't to argue that Mop flies represent crane fly larvae, nor is it to convince you to use them. Instead, recognize that there is lots to be learned from this fly, including the fact that the main material isn't fly-tying specific; instead, it's used daily around us, found on bath mats, children's toys, and more. My first purchase was a set of "cleaning slippers" from Amazon, and once they arrived, my wife Heather thought it would be hilarious to clean our kitchen floor with them when I got home from work. Yeah, it was a little funny. The Mop teaches us to be observant with everything, as the best fly-tying ideas don't always come out of a fly shop.

Tying the Mop Fly

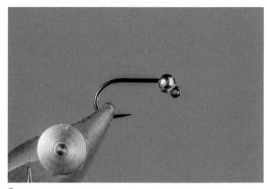

1. Place the slotted bead onto the hook, inserting the point through the smaller hole. Slide the bead into position against the eye, with the slot facing up. Consider oversizing the bead for this pattern, which will help ensure the fly rides lower in the water column.

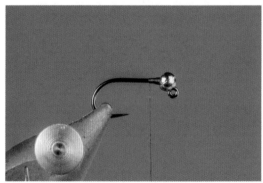

2. Begin with thread wraps jammed tight against the bead, building a thread dam to hold it in position. Continue wrapping slightly rearward, allowing the thread ramp to naturally push the wraps to the shank. Trim the thread tag.

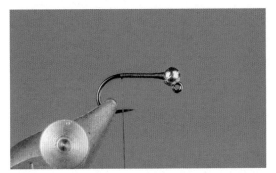

3. Wrap to the rear of the hook, building a thread base on the shank with touching wraps. Spin the bobbin counterclockwise to uncord, which allows the thread to lie smooth on the shank.

4. Measure the Mop material so the tip extends the approximate length as the shank. Trim the butt end so it will lock in place tight against the bead. In this picture, you can see the recommended length in which I tie and fish the Mop fly.

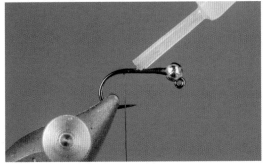

5. Using brushable superglue, soak the entire thread base and all sides of the shank. Take your time and be intentional when using liquid resins, as excess may drip onto the hook point, vise base, or tying bench.

6. Place the Mop material on top of the hook shank, then use your left pointer finger and thumb to press down with gradual pressure so its core fibers are tight and centered to the shank.

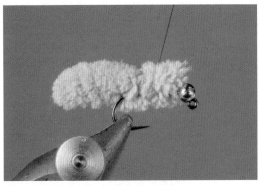

7. Wrap thread forward through the Mop material, making three or four total turns to the eye. Use maximum pressure and wiggle the thread through the Mop fibers, which will help ensure they don't get matted down.

8. Continue winding toward the bead and, once reaching it, add a few additional thread wraps to lock any fibers in place. ***Tying tip:*** While wrapping, your goal is to have the Mop material completely cover the thread base, which is soaked with superglue. As you experiment with different sizes and diameters of Mop material, you'll notice that some types will completely surround the shank, whereas others will rest more on top.

9. Create a short dubbing loop with thread at the bead, then lock it in place with thread wraps both behind and in front of the loop. In this picture, I used a dubbing loop tool, but you can use a needle to split many threads, which is another fast method.

10. Place dubbing into the loop, then slide it close to the hook. Only a sparse amount is required for the collar, and it is easy to overapply. For the Mop, I want to add some sparkle and movement at the front of the pattern, and another material I frequently choose to use inside a dubbing loop is CDC.

11. Begin spinning the dubbing loop, causing the thread and dubbing to form a tightly bunched collection of fibers. While spinning, if your end result looks uneven and scattered, start over and take additional time to place an even amount of dubbing throughout the dubbing loop. This includes ensuring that the dubbing ends are the same length on both sides of the loop.

12. Take your fingernail and scratch the dubbing loop to loosen any trapped fibers. Once complete, you should have a short and even section of dubbing extending ½ inch on each side of the thread.

13. Begin winding the dubbing loop between the Mop and the bead, palmering the dubbing with each wind. Gently stroke the fibers rearward to help prevent them from getting trapped with each successive turn. Keep your total wraps to a minimum. (I made two turns in the one pictured here.)

14. Secure the dubbing loop with your tying thread, locking it in place tight against the bead. Trim the remaining section of dubbing loop.

15. Brush superglue onto the thread prior to completing a whip finish. This is an optional step for many tiers, yet one I recommend to provide this fly (and many others) with extra durability.

16. Whip finish at the head of the fly, taking time to ensure that your thread turns stay between the dubbing and the bead. *Tying tip:* Moisten the dubbing and stroke it rearward prior to the whip finish, which will prevent many fibers from getting trapped by the thread. This technique is preferred when winding straggly fibers forward with touching wraps.

17. Trim the thread tight against the fly, being careful not to cut any dubbing fibers.

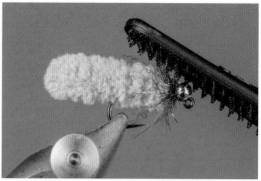

18. To loosen any trapped dubbing fibers, move a dubbing brush back and forth over the collar. My preference is to brush toward the eye first, doing so around all sides of the hook, then stroke the dubbing fibers rearward.

19. In its simplest form, an effective pattern uses only the Mop material sans any dubbing. Recently, more tiers and fly fishers have found ways to incorporate the Mop material into other patterns, and I've seen it added to saltwater, carp, and stillwater flies, too. With a variety of Mop material on the market in different sizes and colors, experiment away!

Sexy Walt's Worm

Though there's not much to this simple pattern, we have lots to examine by first returning to the basics of fly tying, then building in some modern materials and techniques that will enhance your future nymph patterns.

Stepping into the 21st century, our modern-day replacement for a Hare's Ear is the Walt's Worm . . . and here's one with a few extra goodies! Famed Pennsylvania champion fly tier Walt Young developed the Walt's Worm in the 1980s, and the tying steps below illustrate current techniques that can be applied to many types of flies in your box.

When selecting patterns for this book, I was constantly on the lookout for those that shared new ideas and thoughts in fly tying, and I had my reservations with this pattern. In its basic form, the fly can simply be wire with dubbing wrapped over the top, hardly a worthy recipe for a book on tying techniques. However, you'll find some additions that I hope will jump-start your creativity when tying at your bench.

Sexy Walt's Worm

- **HOOK:** #16 Hanak H 450 BL
- **BEAD:** Copper 2.5 mm slotted tungsten
- **THREAD:** Tan UNI 8/0
- **TAG:** Glo-Brite
- **RIBBING:** Pearl UNI Mylar #16
- **BODY:** Hareline Dubbin
- **THORAX:** Bighorn orange Sow-Scud Dubbing
- **HOT SPOT (OPTIONAL):** Fire orange UNI 8/0

The wild brown trout of Central Pennsylvania can be finicky, even when nymphing. This gorgeous fish was no match for a Sexy Walt's, even in a low-water situation. PHOTO BY HEATHER CAMMISA

Let's start the conversation with the addition of a tag, currently called a "blowtorch" on social media, for obvious reasons. Versus imitating a tail, the material screams attractor, and allows you to vary the color based on personal preference . . . yours or the fish's. In some cases, especially for a less substantial rear hot spot, an effective technique is to tie the material "in the round." In a recent tying class, a student accidentally tied a piece of Glo-Brite as a tag *under* the tailing fibers, which looked great, too. When thinking about integrating hot spots into your tying, try to vary the tie-in points and you will be rewarded by the fish.

Referring to the Hare's Ear, another modification is the use of varying ribbing materials versus the standard gold tinsel. UNI Mylar is a great choice, allowing for a more subtle light reflection. Other popular choices include Sulky Metallic Tinsels, which provide great flash and are incredibly strong for their narrow size. Don't forget the use of fine tippet, such as 6X monofilament or fluorocarbon, which makes a nearly invisible material, but can still create that "ribbed" look you're going for. I also know many tiers that will counter-rib a tinsel

with 6X tippet to protect the tinsel from tearing. Though it may seem like overkill, that small addition will allow for increased durability when catching larger quantities of fish. Lots of options to experiment with today!

As the tying draws to an end, there are multiple stopping points, and each depends on one of these questions you may ask yourself:

- In a hurry? Whip finish and go!
- Want a buggier look? Use a piece of Velcro over the dubbed body.
- Fishing spring creeks and want a flatter profile? Brush out the body with a brush, then trim just the top and bottom.
- Prefer multiple hot spots? Grab some fluorescent thread and whip finish directly behind the bead.

The tips and techniques used on this pattern can be applied to many flies in your arsenal, especially those with dubbed bodies. When Walt Young created such a simple pattern, he had no idea it would lead to a multitude of variations . . . nor that it would *ever* be "sexy!"

✓ Tying Tip

Tying with scissors in hand can be difficult for many to master, but as you near the end of this pattern, doing so will save time. The next time you're tying, find a comfortable position for scissors in your right hand (for right-handed tiers) and leave them there for five minutes. Over tying sessions, gradually increase the time, and they will become more comfortable to hold as you are tying. This didn't come naturally for me, but once I switched to a lighter pair of scissors, I found them in my hand more often. Find a comfortable pair that works for you.

✓ Featured Technique

One of the more difficult techniques to master when beginning tying is the application of dubbing. Your goal is to use the thread as the core and spin dubbing material around it, creating a slender strand to wind over the hook shank. Some tips that I can offer include:

- Pull a clump of dubbing from the package and rework the material to loosen the fibers, making them much easier to work with.
- Moisten your fingers to help "grip" the dubbing. If the fibers are still tough to grasp, especially when using synthetic dubbings, apply a tiny amount of dubbing wax to the thread to help start the noodle.
- "Less is more" with dubbing, and once you've taken a pinch, cut the amount in half and start spinning the noodle.
- When looking down the thread toward the bobbin, apply the dubbing by spinning it between your fingers as they move clockwise.
- Always rotate your fingers in the same direction versus back and forth; the latter causes your dubbing to twist then untwist, ruining your chance at a tight strand.
- Once you've formed a tight strand of dubbing and started wrapping, you'll notice that it can be wound over itself to build a tapered body.
- Slightly more advanced: Form a carrot taper of dubbing to the thread by applying less material closer to the fly and more material down the noodle.

✓ Materials to Consider

In the world of dubbing, you have about a million different choices, with more added yearly. When selecting dubbing, I consider how spiky I want the fibers to be and then attempt to use one that matches. For instance, if I want lengthier body fibers, I'll choose something like a SLF Spikey Squirrel blend or Angora goat. For tighter dubbing noodles, there are many materials that I consider favorites, especially beaver fur, superfine dubbing, and hare's ear blends with light amounts of flash. On the other hand, if I want a unique color that I know no one else will have, I may blend a creation of my own, such as dark hare's ear (75 percent) with a fuchsia Antron dubbing (25 percent) using an old coffee bean grinder.

✓ Fishing Suggestion

When fishing patterns like this style, I select the color based on water types. On a spring creek, the fly will have more gray in the body to imitate cress bugs. If fishing freestone rivers and streams, the color preference will be a brown hue, lighter for caddis and darker for certain mayflies. More hot spots get added for fish in off-colored water and deeper pools, whereas I may use fewer when the fish are spooky. These are not absolute rules, but general guidelines offered as a starting point to fly selection.

I recommend fishing the pattern, which is not intended to specifically imitate anything, in varying water types. This includes riffles, pools, along edges and seams, and even in lakes. Depending on the water depth, vary the size of the tungsten bead to help your fly get to the desired depth. When fly fishing moving water with jig nymphs, I prefer to use a Euro nymph setup, allowing a tiny amount of slack in the system which will cause movement in the sighter or tippet to help you detect a strike. When fishing small nymphs in windy conditions or at distance, the use of a dry-dropper setup or strike indicator is preferred.

Tying the Sexy Walt's Worm

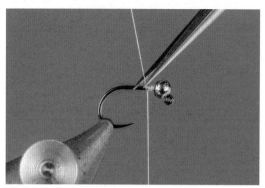

1. Place the slotted bead onto the hook, inserting the point through the smaller hole. Slide the bead into position against the eye, with the slot facing up. Begin with thread wraps jammed against the inside of the bead, building a thread dam tight against the bead so it won't move rearward. Make six touching wraps toward the rear, then trim the tag end of the thread.

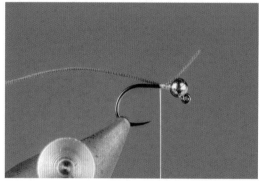

2. Cut a short section of Glo-Brite and secure it with a pinch wrap close to the bead. Take two additional turns of thread, but do not apply maximum tension.

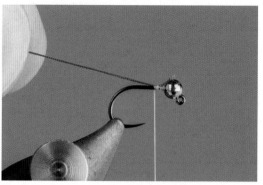

3. Gently pull the Glo-Brite to the left, toward the rear. Since there is little tension on it, the tag end will slide toward the locking wraps. This technique will save the step of trimming, while also reducing the amount of Glo-Brite used per fly.

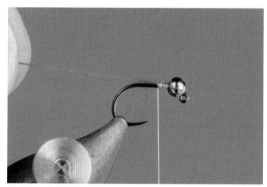

4. Stop pulling the Glo-Brite once it is nearly all under the locking wraps, with the tag end trapped behind the bead. This is a great technique to use for many materials, especially ribbing and shorter fibers that require their entire length to be used.

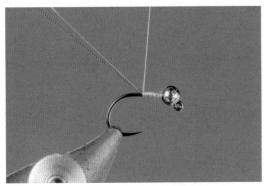

5. Secure the ribbing material with three locking wraps. If the tag extends past the bead, use the previous "sliding method" to gently pull the material in place so the tag rests just inside the bead slot. Keeping materials raised above the hook shank, wrap the thread rearward with touching wraps. Wrap to the hook bend, then place Glo-Brite and ribbing into a vise material clip.

7. Begin wrapping the dubbing noodle forward with touching wraps, encouraging the fly to gently taper as you near the thorax. If your dubbing noodle is the same slender diameter, I recommend that you double back over the previous wraps once reaching the thorax, providing a gradual taper to the pattern.

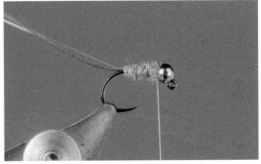

6. Take a pinch of dubbing and noodle it between your finger and thumb onto the thread. Taper the noodle so a sparse amount of dubbing is close to the top of the thread, with a greater amount nearer the bobbin. Slide the dubbing noodle up the thread so its end is nearly touching the hook shank.

8. Wrap the dubbing noodle until approximately one turn before the bead. Remove any remaining dubbing from the thread (and set it aside for your next fly).

9. Counter-rib the tinsel with firm wraps, ensuring even spaces between each wrap. Continue the ribbing to the bead, stopping the last wrap by the slotted opening of the bead.

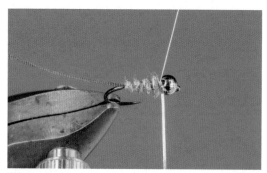

10. Bring the thread to the left underside of the ribbing, then wrap it around the hook shank, locking the ribbing in place. Make multiple thread wraps to secure the ribbing to the shank.

11. Advance the thread in front of the ribbing and make two wraps, then cut the remaining ribbing close to the hook shank. *Tying tip:* Keep your scissors in hand, as they're used in the next step. Many tiers keep their scissors in a comfortable position in hand throughout the entire tying process, reducing the number of overall steps.

12. With scissors still in hand, rest them against the hook bend and angled slightly forward. Trim the Glo-Brite to the desired length in one snip. The length of the hot spot varies based on

personal preference; I prefer mine to extend just inside the hook bend.

13. Apply desired head cement or superglue to the thread with a brush, then whip finish and trim the tag end of the thread. Need more reinforcement for the thread? Dip your bodkin into head cement and gently apply it around the thread collar. This can also be accomplished with a slight amount of thin UV resin, which should then be cured with a UV light.

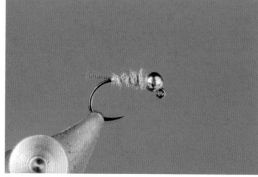

14. At this point, many would call this fly complete. You can remove the hook from the vise and fish this pattern, and others like it, with confidence. But this is a learning book full of additional techniques, so let's keep going!

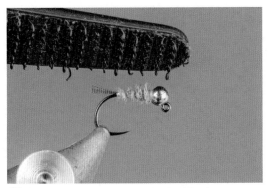

15. For an overall buggier dubbed body, use a dubbing brush swept back and forth. When attempting to imitate flatter nymphs, I will only use the brush on the left and right sides of the pattern and trim the top and bottom with scissors. If I want a completely buggy look, then I will brush around the entire body of the fly.

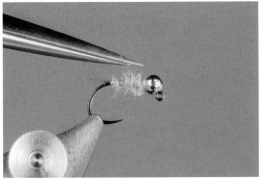

16. Use your scissors to trim away any excessively long fibers, though I know *many* fly fishers who prefer to let the fibers remain. Long fibers will encourage more movement in the water, but will give the appearance of a thicker body, too. This can also prevent the fly from sinking quickly in fast-moving water . . . which means it's a great technique to use when fishing in slower pools.

17. Grab a preferred fluorescent thread, tie it in with a few wraps, then trim the tag. Continue wrapping until the space between the thorax and the bead is filled, then build it up to the approximate level of the ribbing. Should you have used a fluorescent thread all along? Depending on the overall brightness of the thread, some will show through the dubbing (which isn't always a bad thing). Experiment with your threads, as using one through the entire process saves tying time and steps.

18. You can vary the dubbing type and color on this style of generic nymph to imitate insects in your local waterways. For example, a favorite stream of mine has a significant Blue-Winged Olive population, and fish love nymphs of all sizes . . . as long as there is some brown-olive dubbing on whatever I'm fishing! ***Tying tip:*** Need more durability? Add an additional ribbing of 6X fluorocarbon, which will protect the mylar tinsel ribbing and thread underneath. Now we're talking about a formidable fly!

Perdigon

The Perdigon is a slender nymph pattern that gets to the bottom in a hurry. The contrast created with its dark wing case covering a hot spot guarantees to elicit a reaction from many fish, especially in faster water.

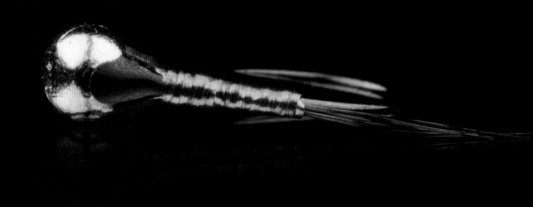

Occasionally there is a fly-tying pattern that takes our sport by storm, and the Perdigon absolutely qualifies as one of those! This slender-bodied nymph with its distinctive look almost appears oversimplified, but the few materials contribute to the overall effectiveness. Related to this, the Perdigon features two distinct characteristics that separate it from the pack.

First, you'll notice a resin-coated body, which is intended to help it slice through fast currents to catch those fish that reside near the bottom of the water column. Next, the thorax is mainly brightly colored thread, contrasting with its trademark black wing case. UV resin is applied over everything (sans the tail and a portion of the bead), securing the materials and giving a nice glossy

Perdigon

- **HOOK:** #14 Hanak H 450 BL
- **BEAD:** Silver 3.0 mm slotted tungsten
- **THREAD:** Olive Semperfli 12/0 Waxed
- **TAIL:** Medium pardo CDL
- **BODY:** Olive Krystal Flash
- **THORAX:** Fire orange UNI 6/0
- **WING CASE:** Black Solarez Bone Dry
- **UV RESIN:** Solarez Bone Dry

finish. Sounds easy enough, and we've benefited from modern materials to help achieve this in a short amount of time.

The Perdigon's tying procedures have evolved even since this book was in its early stages, showing how the fly-tying market adjusts much quicker today to popular patterns. Traditionally (if that word can be allowed with this one!), the Perdigon's black wing case was formed by a variety of materials, including nail polish or even a black Sharpie, which both have their shortcomings. The pattern was typically tied up to that point, then placed on a drying rack with others waiting for their nail polish wing case to dry before receiving the final resin coating. Since the advent of colored UV resins in fly tying, very little wait time occurs, as curing with a UV light takes less than 15 seconds. The original Perdigon selected for this book was to feature nail polish, but once I got my hands on black UV resin, it was an easy switch to make.

Let's examine this fly closer, as many of its trademark techniques can be connected to other slender patterns. Starting with the tail, I have been asked *many* times over the last few years if I count the number of fibers, and the answer is, "Yes and no!" Taking a step back from the Perdigon, the key to patterns like this is their ability to cut through the water, and more tail fibers will slow it down. Yes, we're talking milliseconds, but time impacts the effectiveness in fast water. So what's the ideal number of tail fibers? Three to five seems to be my magic range depending on the hook size, though I only count that number occasionally. Moving forward, let's examine the body, or lack thereof!

When the Perdigon craze hit social media, I was intrigued and surprised by the simplicity of the various patterns, but soon realized that concise tying was required to achieve the correct proportions of this pattern. From touching wraps and selecting thin materials, my guiding principle to you is this mantra: "Less is more." The selection of flashy body materials is something I lean toward with these patterns, knowing they'll be fished in fast-moving water. Giving the fish that extra glint may be the difference maker between this fly and one with more subtle colors; hence, even when I'm fishing a black-bodied Perdigon, my starting

material typically features fine tinsel, and favorites include Semperfli Perdigon Body or Hends Perdigones Body. Be careful when wrapping with certain materials, as overstretching them may create a reverse taper. Once coated with a resin, that flash is ready to roll!

Speaking of resins, fall back to our mantra and apply only one or two thin coats. When teaching tying classes, this is one of the biggest mistakes I see with Perdigons, as tiers continually build tapers and excessive resin layers, in turn hindering the overall future performance of the pattern. Two layers make this fly bulletproof, with extra ones offering little in the design department. If there's any excess after applying resin, attempt to dab it with a piece of paper, such as a Post-it Note, as using a tissue or paper towel can leave unwanted particles. Otherwise, once the body appears to be wet, I hit it with my UV light, curing the resin.

One of the key characteristics that originally drew me to this pattern is the thorax, which is different from most nymphs I've ever tied. Simply consisting of fluorescent thread or floss, the thorax mainly plays the transition role from body to bead, creating a fluid taper. There are times I'll experiment with the colors and types of threads used for this, though I recommend a larger-diameter thread, such as 6/0 or 140 denier, which allows for faster buildup of the section. The thorax is a critical piece, though the most important component is what lies on top of it—that sexy black wing case.

To me, nothing says Perdigon like that wing case contrasting with a brightly colored thorax. Prior to using black Solarez Bone Dry, I experimented with various nail polishes, trying to determine which one would dry the fastest. The drying factor meant that the fly had to be tied in stages, removing each from the vise while waiting for the wing case to dry, then curing them, assembly-line style. As modern UV resins have advanced, there are also more colors to choose from, driving the indecisive tier crazy! Don't let that deter you, and be sure to examine today's choices and consider how they can enhance some of your favorite patterns.

The final thought I have about this wing case is that many suggest it should be tied on the inside

When using UV resins on many patterns, I notice that tiers have a tendency to apply an excessive amount. The first step in using resin is to dab the brush against its container, allowing the excess to run back inside.

of the shank, as the jig hook will cause this fly to invert in the water. Yes, this style pattern when paired with a slotted bead tends to ride hook-point up, but as it tumbles through fast water, fish have the opportunity to see it at various angles, and the key is to create a visible contrast between the wing case and thorax. If the fish start questioning why our wing case doesn't precisely match the natural, we are all in trouble!

✔ Tying Tip

Not all UV resins are created equal! Some require additional steps after curing; hence make sure you read the directions on the one you choose. A common question I receive via my YouTube channel relates to curing time: If your flies are still sticky after an extended period after using your light, I recommend changing the batteries. Once they get to a reduced charge, you'll notice a significant increase in curing time. Consider using a rechargeable light or batteries if you decide to integrate UV resins into your tying arsenal.

✔ Featured Technique

Applying UV resin can be tricky, especially when using a runny product, commonly referred to as "thin" in the world of resins. After unscrewing the applicator tip, I hold it above the bottle and allow some initial runoff, then stroke the brush against the inner neck. For the Solarez applicator, above the brush is a hard plastic connector that tends to hold additional unwanted resin; thus I'll place that connector directly against the inner lid, again allowing it to run back into the container. Once sure I only have the bristles wet, I'll begin application to the body, working around from top to bottom. If I ever notice excess, it will typically be a drop that has formed on the fly's underside and is easily removable by dabbing with a small Post-it Note.

✔ Materials to Consider

The choices today for Perdigon body materials are truly unlimited! When thinking about thin and brightly colored materials, a quick glance at your local fly shop pegboards and thread section will overwhelm you with all the colors of holographic tinsel, GSP threads, flosses, and synthetic body quills, all of which have a place on the Perdigon. Take this one step further and consider ribbing the body prior to curing, which you can do to create a contrasting color from the body, or to simply show body segmentation.

✔ Fishing Suggestion

Slender nymphs have risen in popularity because of their ability to cut through fast currents, allowing us to fish in spots that typically required bigger flies, beads, and lots of weight attached to the tippet. This pattern is a favorite in deep and fast riffles, and I'll turn to it in situations where I have a short amount of time to get deep in a hurry. Upon reading the water, I'll turn to Perdigon-style flies in faster currents versus slow pools, then encourage a faster sink rate with 5X to 7X tippet. Does this mean Perdigons have no business in shallow water? Of course not, and that's where you'll find some of my smaller-sized ones being put to use.

Tying the Perdigon

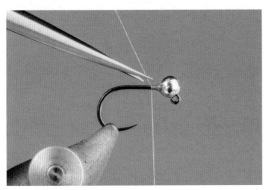

1. Place the slotted bead onto the hook, inserting the point through the smaller hole. Slide the bead into position against the eye, with the slot facing up. Begin with thread wraps jammed against the inside of the bead, building a thread dam tight against the bead so it won't move rearward. Make six touching wraps toward the rear, then trim the tag end of the thread.

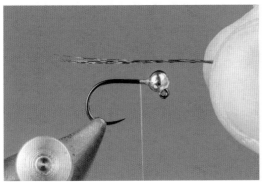

2. Trim or pull three to five fibers of Coq de Leon from a feather and measure against the hook shank. The tips should extend past the bend of the hook approximately the same length as the body. After measuring, transfer the fibers to your left hand, keeping the tips pointing away from the hook rear. Secure fibers with three pinch wraps tight against the hook shank.

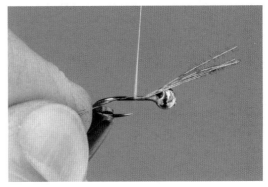

3. Begin wrapping touching wraps rearward with the thread. While winding, gently pull the tips of the CDL fibers toward you as the wraps progress. Doing so will encourage the thread to gently twist the fibers into position along the top of the hook shank. Continue wrapping toward the rear of the hook.

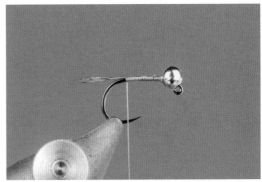

4. Stop wrapping the thread at the point where the hook begins to bend down, then trim the butt ends of the CDL. *Tying tip:* There are times when I prefer the tails to sit at an angle facing up from the hook, and it's then that I will occasionally place a thread wrap between the CDL and the hook, encouraging that slight angle upward.

5. Place three strands of Krystal Flash against the top of the hook shank, with their butt ends nearly touching the bead. Using a pinch wrap, secure the Krystal Flash, then continue wrapping forward with touching wraps.

6. Build a slight carrot taper as you bring the thread toward the bead. Once you're satisfied with the taper, leave your thread waiting one wrap from the bead. To build a full taper, I prefer to fully wrap the thread to the bead, then return approximately three-fourths of the way to the rear. Then I wrap to the thorax, and back halfway to the rear. This back-and-forth wrapping will allow the taper to progress slightly. Remember that a narrow taper will allow the fly to sink faster, and vice versa.

7. Wind the Krystal Flash forward with touching wraps, and ensure that the thread underneath is completely covered while doing so. Wind completely forward until reaching the front of the thorax, then secure the material tight against the Krystal Flash. Be gentle while winding Krystal Flash, as pulling the material will stretch it, causing its coloration to lighten significantly.

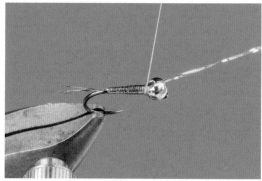

8. When tying off, encourage the Krystal Flash to rest in the bead's slot, then trim the tag so the end fibers disappear into that gap.

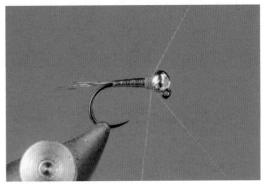

9. Whip finish the thread tight against the bead and trim it close to the body. Adding head cement directly to the thread is optional but not recommended, as you'll be covering the thorax with a hot spot and UV resin to seal in later steps.

11. Whip finish the hot spot tight against the bead, which will help fill in any remaining wraps when completing the taper. Be sure to trim the thread tag close to the body; if it is sticking up during the UV resin application, the thread will cause an unneeded bump in the thorax. Another tip is to gently wet any remaining thread tag and encourage it to sit flush with the thorax before adding resin.

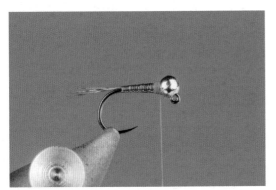

10. Tie in a fluorescent-colored thread and build an aggressive taper toward the bead within the thorax. Continue wrapping until you have a few remaining turns close to the bead; save the rest of the taper for the whip finish, which will finalize the end of the taper. Use a thread or floss with a larger diameter for the hot spot, as it will build up quickly and reduce your overall time.

12. Apply a thin coat of Solarez Bone Dry UV resin, taking time to cover the entire body and hot spot completely around the hook. If too much resin is applied and you notice some pooling on the underside, touch it lightly with the side of a Post-it Note to absorb the excess away.

13. Be sure to completely apply resin to all openings, including within the bead's slot. *Note:* Additional coats are optional, as they will increase the fly's durability, but also add increased resistance. This causes the pattern to sink slower, thus potentially taking away from the slender pattern's effectiveness.

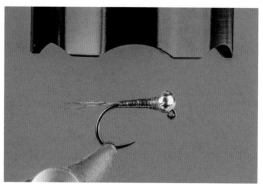

14. Cure the resin with a UV light, shining it evenly on all sides of the fly. Base the overall time on manufacturer's recommendations; if curing is slow, be sure to check the batteries in your UV light.

15. Prepare the brush for the black Solarez Bone Dry UV resin by dipping the fibers into the resin, then gently dab the sides against the inside neck of the bottle. The brush is ready when the resin appears even from top to bottom.

16. Apply a thin coat of black Solarez Bone Dry UV resin to the top of the thorax, ensuring that the resin is only applied to the top half of that section. I prefer to dab a tiny amount behind the bead, then work toward the slot. *Tying tip:* Less is more when first learning to apply UV resins, and adding multiple coats of extremely thin layers will produce better outcomes versus applying an excessive amount.

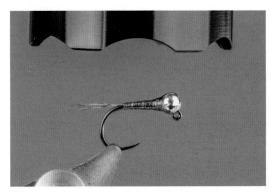

17. Cure the resin with the UV light, shining it especially over the newly created wing case. Be sure to allow this to completely cure, as another coat will be applied directly on top, which may prevent the inner coat from curing.

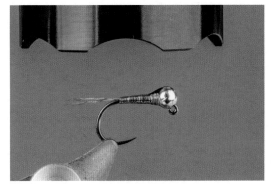

19. Cure the resin, taking time to shine the UV light over the entire Perdigon and the UV-covered sections. If at this point you notice any portions missing resin, take a moment to reapply and cure again.

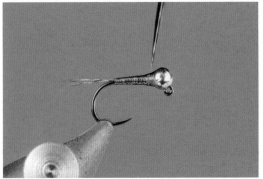

18. Apply a second coat of black Solarez Bone Dry, with your goal being to create a top taper that begins at the rear of the orange hot spot and connects with the top of the bead. This can be accomplished with one or two coats, depending on the size of the hook. When using other materials for the wing case, such as nail polish, it's recommended to apply an additional coat of clear UV resin over top to seal everything.

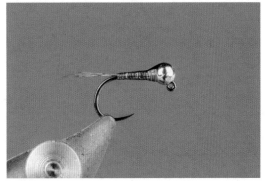

20. This Spanish nymph has endless variation possibilities, including altering both the body and the thorax hot spot. Previous iterations used black nail polish or a Sharpie for the wing case, though the presence of UV resin in a variety of colors has made this step incredibly easy, while ensuring overall durability of the fly. When examining your pattern, be sure to verify that it was tied *thin*, and if not, be intentional with your thread and material wraps and the application of UV resins, in the future.

Beach Body Stone

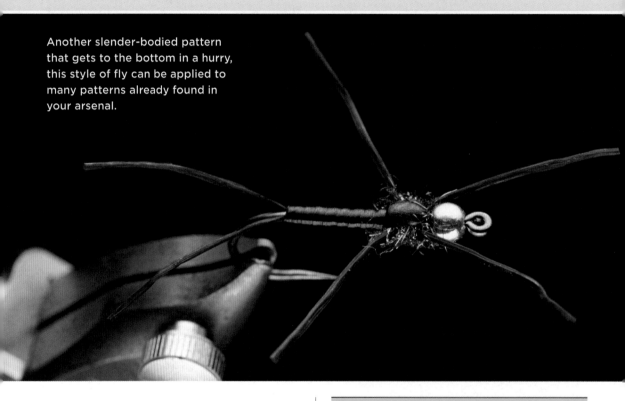

Another slender-bodied pattern that gets to the bottom in a hurry, this style of fly can be applied to many patterns already found in your arsenal.

When contemplating various stoneflies to include in this book, the emphasis was squarely on selecting a pattern with a variety of fly-tying techniques, and this Euro-style pattern certainly involves more than I anticipated! A creation of Fly Fishing Team USA member Josh Miller, the recipe appears straightforward and simple, yet achieving the slender finished product requires you to maintain focus while being intentional with your thread wraps, which is where our discussion starts.

Gel-spun threads (GSP) are incredibly strong, especially when their thinner diameters are taken into consideration. My first exposure to this was watching Mike Romanowski tie a gorgeous Comparadun and place a tremendous amount of

Beach Body Stone

- **HOOK:** #8 Hanak H 970 BL
- **BEAD:** Silver 3.5 mm tungsten
- **WIRE:** .015 mm (12–15 wraps)
- **THREAD:** Black Semperfli Nano Silk 18/0
- **TAIL AND LEGS:** Black Life Flex
- **BODY AND WING CASE:** Black scud back
- **THORAX:** Black and purple Ice Dub, 50/50 blend

pressure on his thread, way too much in my opinion . . . but it didn't snap! The wing was formed perfectly, and I immediately had to try the new threads. After a bit of experimenting, GSP threads became a part of my tying arsenal, especially when traveling. Related to this pattern, after finishing the wire wraps, there are times when I'll simply increase the thread pressure and slice through the wire tag, saving a tying step and time.

As my great-uncle John loves to say, "What you do now, you don't do later," and I recommend you take the time to ensure a smooth thread transition between the wire and body. Wrapping the scud back will go a lot easier, and that tapered body gives the cigar-shaped appearance that fly tiers (and fish) have come to love and expect on many nymph patterns. As this pattern is continued, you'll notice that pressure is applied to both the Life Flex and scud back materials when locking them in place; that stretching pulls them tight,

encouraging a thinner body. What else would you expect with a fly named after a diet program?

Continuing with "stretchy" materials, there are many techniques to assist in getting them to either stay in place or add minimal bulk. The tie-in spot of the legs and scud back is intended to help naturally taper the pattern, whereas tying both at the rear of the body would result in the dreaded "bump" at the end of the hook. In the tying pictures, notice how my finger and thumb are placed inside the Life Flex when creating the tail, encouraging it to stretch, while my thread is wrapped toward the bend, helping to keep the thin profile. My finger placement also places both pieces on the sides of the hook, which ensures the tail separation after arriving at the last rearward wrap. Push yourself to choose tie-in locations that encourage fewer thread wraps; you will be rewarded with slender-bodied flies that get to the strike zone faster.

During a high-water event in Iceland caused by glacial melt, the best way to get patterns to the bottom was to use a heavy stonefly on the point. My reward was this gorgeous Arctic char, easily the most beautiful fish I caught during my last trip there. COURTESY OF BLACK MOUNTAIN CINEMA

Occasionally, I like to share a little "behind the scenes" in the world of YouTube, and here's a story that few know. Once Josh had created this pattern, he decided that sharing it with others was the right thing to do. It had accounted for lots of fish caught and integrates some great tying techniques. To get the pattern out to the public, he chose my YouTube channel, and we met at my house for the recording. Once ready to begin filming, I asked him to tell me about the fly so I could understand a few of the nuances, then asked for its name. The look on Josh's face told me he didn't have one, which levied some pressure on us, as we were moments away from starting. As an impromptu brainstorming session began, Josh reminded me that the key to this stonefly was its slender appearance, much more so than most others fished at that time. In 2016 the Beachbody diet was popular in many realms of society, and I tossed Josh the idea of a "Beach Body Stone" to imply a fly that's been on a diet. He liked the name and we pressed "record"—and now you know how some fly names are created!

✓ Tying Tip

The main advantage of GSP is using it in situations that require little thread buildup, while being able to apply maximum tying pressure. When traveling, a dark and light spool of GSP 18/0 will tie nearly 90 percent of the flies needed, so it has reduced the amount of materials I take on the road. A trade-off I've experienced is that these threads tend to almost have a sheen to them and unstrand easier than most, but I can easily see past that when having the ability to tie patterns that range from size 24 to 6 without changing a spool. To help with cutting GSP threads, be sure to hold the tag end taut before snipping. A final consideration to note is that GSP threads may dull scissors when snipped directly. It's best to cut farther away from the scissor tips or even choose an alternative tool, such as a razor blade, when using GSP.

✓ Featured Technique

Blending dubbing on your own can be tricky, which is why many tiers prefer to purchase prepackaged dubbings. From my perspective, I prefer blending, as it ensures more color consistency while also making the dubbing used on my patterns unique (and possibly a shade the fish haven't seen). To blend dubbing, I use a "retired" coffee bean grinder, pulsing it for anywhere between 10 and 20 seconds. If you choose to go this route, here are a few pieces of advice:

- Fast is fine, but accuracy is everything. There is nothing worse than creating an almost magical dubbing blend that has the perfect color and just the right amount of flash . . . then forgetting what you mixed together! As I add material to the grinder, I keep track of everything, then write the finished blend in Sharpie on the material baggie.
- Just a pinch. I prefer to add materials a pinch at a time, and the key is to make sure the amount is consistent regardless of material type. Doing so helps ensure an accurate count at the end, making the blend replicable.
- Creativity is king. The first two tips make the blending process come across as black and white, but there are times when I start adding materials just to see what will happen. Many times . . . epic failure! But occasionally I'll stumble across a new dubbing blend that fits perfectly with a pattern, such as the time I started with a 50/50 blend of black and purple Krystal Flash for this one, then added pinches of red until I found the perfect mixture. I'd let you know exactly how many pinches, but that's my secret blend. Good luck finding yours.

✓ Materials to Consider

In fly tying, varying materials is accepted and encouraged, which sometimes is easier said than done. For instance, when Josh mentioned the Life Flex legs on this pattern, I only had the smaller size, which had a thinner overall diameter. How did I adjust? Simple, by purchasing another size! Reflecting on flies with rubber legs, I prefer to maintain various diameters, especially based on the water type I'm fishing. Smaller diameters require little current to ensure adequate movement; they're selected when fishing slower sections of rivers, and even in stillwater situations. A larger-diameter selection is useful in faster currents, as the water will easily bend the legs back, yet their increased diameter encourages them to spring forward, causing those vibrations we know many fish can't resist. What I'm politely trying to say is simple: The next time you watch one of my videos and are missing the specific material I've suggested, don't worry! More than likely what you have on hand will work . . . but if you're like me, you'll probably still find a way to get the recommended one.

✓ Fishing Suggestion

This pattern gets deep in a hurry! Keep that in mind, as its use in shallow, slow water can results in losing way too many flies. Fish heavier nymphs and stoneflies in a variety of other situations, which can include:

- Fast runs. Fishing deeper riffles can be difficult, especially with an indicator, as the faster surface current will attempt to pull your line down-stream at a greater rate than the fly should be traveling on the bottom of the stream. To com-bat this, I recommend lengthening your tippet to extend through the surface, similar to a Euro nymph setup. Using a longer tippet will slow things down, keeping heavy patterns in the strike zone and giving you a better connection.
- Deep pools. Fishing heavily weighted flies in these sections will put your pattern closer to the fish, giving them an opportunity to say yes! In many instances, I will use a Beach Body Stone as my point fly, then about 20 inches up the tippet have a smaller dropper fly. The heavier pattern helps get my dropper deeper in the pool, something that can be difficult to accomplish with smaller nymphs. If the fish are feeding closer to the bottom, the point fly is there; those fish actively chasing nymphs will rise through the column, finding your dropper fly to intercept.
- Pocket Water. Finishing up this book, I was talking with Josh and he let me in on a guiding secret with the Beach Body and similar stonefly patterns: Steelhead love them in pocket water! Sizing the fly down to a #12 hook, Josh makes

slight adjustments to a Euro nymph leader, most notably decreasing the tippet length to less than 4 feet. Sighter material is used directly above the tippet, allowing for quick strike detection, as there's a shorter overall connection with this rig. When using flies in that size range, tippet size is recommended around 5x or 6x for trout and 3x or 4x for larger fish like steelhead.

Tying the Beach Body Stone

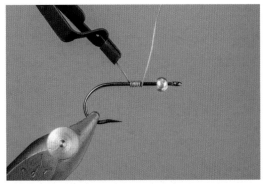

1. Place the bead onto the hook by inserting the hook point through the smaller hole. Slide the bead onto the shank, then secure the hook into the vise, ensuring the shank is parallel to the ground. Make 12 to 15 turns of wire. To prevent the wire from spinning, hold the tag end with hackle pliers between your left pointer finger and thumb, then wrap the right side of the wire toward the bead.

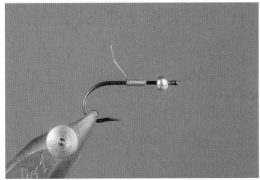

2. Holding the left tag of wire secure, pull straight up with your right hand, tearing off the right tag end of wire cleanly at the shank.

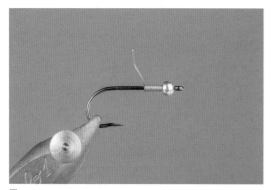

3. Slide the wire to the right, jamming it underneath the bead. This movement should also push the bead tight against the hook eye.

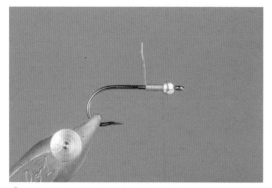

4. Lock in thread to the left side of the wire, creating a shallow dam against it. Wrap the thread through the wire, toward the bead. The thread will naturally lie between the wire turns.

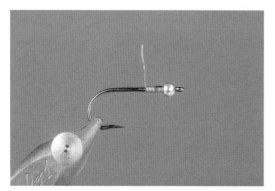

5. Wrap back toward the wire tag, building up thread over the wire to secure the wire to the hook shank. You'll notice that not all the wraps are falling between the wire at this stage. *Tying tip:* Using heavier thread will allow for more

pressure and will rest on top of the wire, but also build up the overall body faster. The use of GSP threads are ideal for this pattern and others, as they have a much higher breaking strength when compared to the diameter of their polyester counterparts.

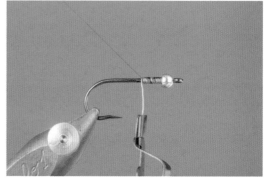

6. Pull the wire tag down over the hook shank and place the last thread wrap over it. Use hackle pliers if you have difficulty gripping the wire with your fingers.

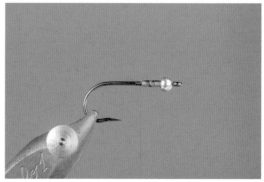

7. Pull down on the thread with slight pressure, evenly cutting the wire off. Other options include using fingernail clippers or scissors to trim the wire. Using GSP thread saves a step, as it is very durable and can slice through wire.

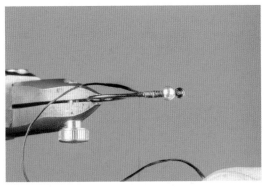

8. Tie in legs on one side of the shank, then allow them to loop around behind the bend of the hook. You're encouraging them to make a circle so you can use fewer materials and steps throughout the tying.

9. Tie the other side of the rubber leg on the near side of the hook, allowing the rubber leg to circle around the rear of the hook. When locking it in place with thread, be intentional and secure the very end of the rubber leg. Scissors won't be needed during this step.

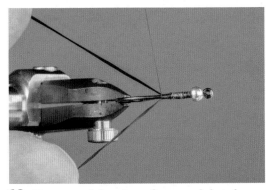

10. Place your left pointer finger and thumb inside the rubber leg loop to keep them apart,

then wrap the thread (holding the bobbin with your right hand) toward the hook's bend with touching wraps.

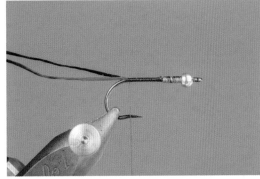

11. Continue wrapping until reaching the bend of the hook. While wrapping, maintain pressure inside the rubber leg loop with your left hand, keeping the rubber legs spread apart so they are both locked in place on their respective side of the hook.

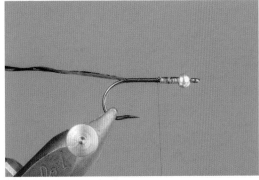

12. Return the thread to the base of the wire with touching wraps, maintaining a slender profile throughout the abdomen.

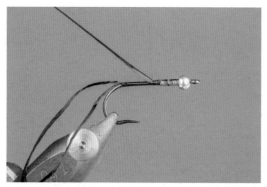

13. Secure scud back material with pinch wraps, making sure it is tight against the wire. Begin making touching wraps toward the hook bend, securing the scud back material to the shank.

14. As you wind the thread rearward, hold the scud back material at a slight angle above the hook. Keeping this material tight will add less bulk to the overall profile of the fly and ensure that the material is secured to the top of the hook shank.

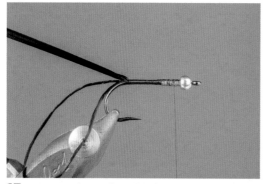

15. Stop winding when the thread reaches the rear of the hook shank, just before it begins to

bend. Return the thread to the thorax, smoothing the transitional taper between the wire and the body. Taking extra time during this step ensures that the taper will make the overall profile of the fly appear seamless.

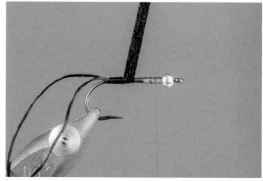

16. Wrap the scud back toward the thorax, pulling it tight with each turn to ensure a thin profile. The first wrap is the most critical; make sure you wrap 360 degrees around the hook so there are no gaps at the rear of the abdomen, then begin winding forward.

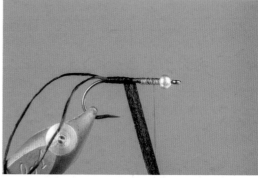

17. Continue winding scud back material forward, with each wrap slightly covering the end of the previous one. Maintain an even pressure on the material, especially as you transition over the wire.

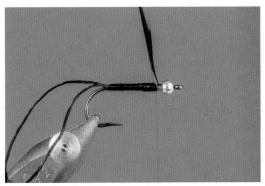

18. Once you reach the middle of the thorax, lock the scud back material in place with multiple thread wraps. Ensure that the tag end of the material is located on top of the hook shank.

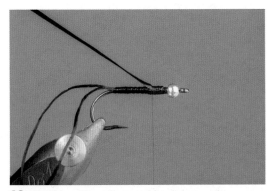

19. While pulling the scud back tag end toward the rear of the hook, wind the thread toward the hook's bend, securing the scud material to the hook shank. Stop winding when the thread reaches the beginning of the thorax.

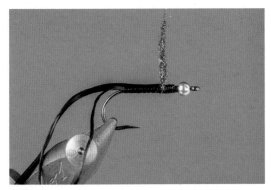

20. Take a pinch of dubbing and noodle it between your finger and thumb onto the thread

so it's nearly touching the thorax. Begin winding the dubbing noodle toward the hook eye.

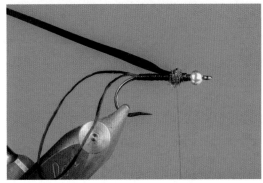

21. Make about two turns forward, approximately one-third of the way into the thorax. Stop wrapping, then pick up another rubber leg.

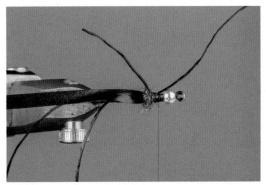

22. Secure the leg on the far side of the hook, with the left tag end pointing away from the hook at a slight angle. Take three thread wraps forward, encouraging the leg material to form a V on that side of the hook.

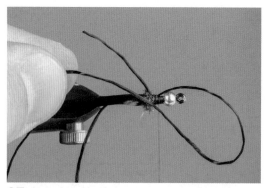

23. Make a large loop with the leg material, going around the front of the hook eye. This loop will form a pair of legs, so be sure to leave it with enough rubber leg material.

24. Secure the rubber leg material on the near side of the hook with thread wraps. At this point, you can enlarge or reduce the rubber leg loop, based on how much material you will need for the legs.

25. Wrap the thread toward the dubbed section, encouraging the legs on the near side to form a V. Some rubber legs simply don't want to go in the desired direction. For those, I pull on the tag end of the material, then secure the material closest to the hook.

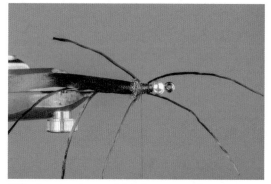

26. Cut the front rubber leg material in the center, allowing both legs to open toward the sides of the hook.

27. Apply a dubbing noodle to the thread, then wrap it forward toward the eye of the hook. Place one wrap tight against the rear of the rubber legs.

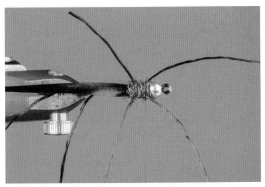

28. Continue wrapping forward, placing another wrap in front of the rubber legs. Continue wrapping the dubbing up to the bead, then remove any excess dubbing from the thread. Be sure to place a wrap both directly behind and in front of the rubber legs, as you want them to remain apart and form an X when viewed from the top.

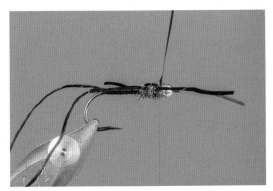

29. Stretch the scud back material forward, holding it tight over the thorax. Secure in place with multiple thread wraps, then hold the material perpendicular to the hook shank with pressure.

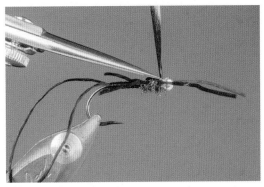

30. While holding the material tight, trim the scud back close to the thread wraps. Notice how the material stretches when held; once cut, it will retreat into the gap between the dubbing and the bead.

31. Pull back the front rubber legs, then whip finish behind the bead. If you dislike seeing thread wraps, apply a sparse amount of dubbing first, then whip finish for a clean look. Fish, however, don't seem to mind.

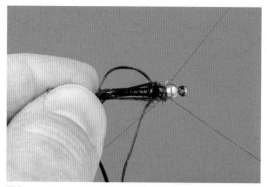

32. The whip finish view from the top. Note how the dubbing is nearly touching the bead and the scud back tag end has nearly disappeared once the pressure was released.

33. Trim the thread, then use a dubbing brush to release any trapped fibers on the sides of the thorax.

34. Continue with the dubbing brush, picking apart fibers on the bottom of the thorax. Once brushed out, the fibers should appear straggly, yet completely fill the thorax underside.

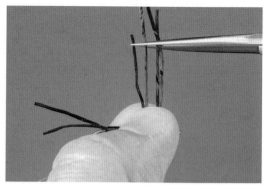

35. Cut all the legs the same length as the body, adjusting based on your own experiences on the water. To get the legs all the same length, prop the front ones together by pushing up with the fingers and thumb on your left hand. To measure the rear rubber legs, simply fold them over the body of the fly and cut.

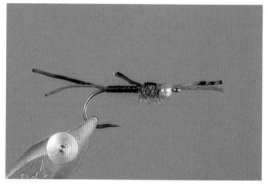

36. Trim the excess dubbing fibers so they are all the same length. ***Tying tip:*** For many subsurface patterns, I prefer trimming dubbing fibers tight to the body, which allows the fly to sink at a faster rate. But in this case, the added wire and tungsten bead help the fly penetrate quickly, and the longer fibers will encourage more movement in the water.

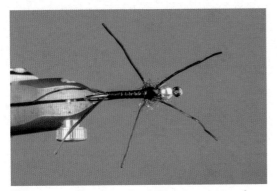

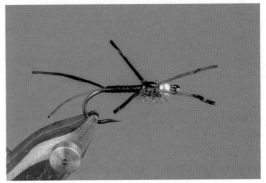

37. Top profile of the finished pattern. Note how the front rubber legs are cocked slightly forward, which will encourage additional movement in the water. As they fight the current, the rubber legs will be pushed back, then attempt to return to this starting position.

38. Be very intentional with minimal material use when tying this pattern, as the slender body really helps the fly sink quickly. When varying the fly, go with thorax colors that you have confidence in, including both subtle ones and those that scream, "Here I come!" The former colors include black and brown, with the latter encompassing hot pink, orange, and red. Finally, examine the style and creativity that is exhibited in this loose representation of a stonefly, and attempt to incorporate this into other flies known to reside in your favorite waters.

STREAMERS

Articulated Streamer

Many tiers shy away from this style of pattern, but understanding the basics of articulated streamers will add another dimension to your tying . . . along with the opportunity to catch some very large fish!

Of all the fly-tying concepts I've learned in recent years, I was most resistant to those surrounding articulated streamers. Sure, some of the largest fish were caught with these patterns (that's 100 percent true), but seeing such uniquely named creations with *tons* of fly-tying materials just didn't make sense. It almost seemed that fly tiers were racing to see who could add the most materials or achieve the longest pattern. Finally, I gave in and tied some basic articulated patterns; immediately I was rewarded with a gorgeous wild brown trout from a stream typically known to produce much smaller fish. Time to learn more about this style of fly!

There are many important facets to these patterns, with the most critical one being movement.

What I learned is that simply adding a hinge point within the streamer doesn't ensure movement; it's just an extension to elongate everything. On the water, the angler absolutely contributes to this through various types of retrievals and movement of the rod tip, but everything starts back at the tying vise . . . so this is where you come in. Selecting materials that move in the water, such as marabou, rubber legs, and schlappen, is part of the process, and then there's another learning curve when it comes to the connector. Yeah, I can see why I was resistant at first.

Right after examining the recipe, I'm guessing some questions popped into your mind. First, what is up with 18/0 thread?! We've discussed GSP threads a lot throughout this book, and I wanted

Look closely at this articulated streamer, and what do you notice? The rear hook is MIA, broken off during a fight with a large fish. During a pressured situation, I chose to simply tie on the pattern and never tested the connection prior to casting. Be sure to take an extra few seconds to check your articulated patterns, as we have to take advantage of the areas of fly tying and fly fishing that we can control.

Articulated Streamer

REAR HOOK
- **HOOK:** #8 Hanak 970 BL
- **THREAD:** Black Semperfli Nano Silk 18/0
- **TAIL:** Burnt orange schlappen
- **BODY:** Hot orange/black EP Brush, 3 inches
- **LEGS:** Amber barred rubber legs

FRONT HOOK
- **HOOK:** #6 Hanak 970 BL
- **CONNECTOR:** Bright Beadalon 19 strand, .018 inch
- **BEADS:** Orange 3-D beads
- **THREAD:** Black Semperfli Nano Silk 18/0
- **TAIL:** Burnt orange barred marabou
- **BODY:** Hot orange/black EP Brush, 3 inches
- **HEAD:** Brown sculpin helmet, small
- **UV:** Solarez Bone Dry

to take an opportunity to illustrate the versatility they offer you, with 18/0's breaking strength being 2.51 pounds. Other sizes of Nano Silk that I commonly use on streamers include 6/0 and 12/0. Were you also wondering about the two different-sized hooks? Because the head typically takes up much of the hook shank, using a larger hook size in front helps create similar body proportions between the two. That nugget of information, shared with me by streamer specialist Gunnar Brammer, is a small detail that makes your pattern that much more effective. Gunnar typically fishes for much larger species, including muskie and pike, so his initial reply when seeing the recipe was not surprising: "Those are some SMALL hooks and THREAD man!"

Sticking with the head, you also have the option of allowing the hook point to ride either down or up, and that tends to be a personal decision

This large brown trout was fooled in a stillwater situation by an articulated streamer tied to mimic local baitfish in the area. Matching size and color is a start, but adding articulation to a streamer pattern gives it an immediate enhancement. COURTESY OF BLACK MOUNTAIN CINEMA

to make. For instance, when watching YouTube videos by Kelly Galloup, I've noticed that he tends to encourage his hook point to ride facing down. Knowing that he recommends fishing a sinking line with streamers, my initial thoughts centered around this fly snagging the bottom, but there's more to the story. One of the more popular methods Kelly uses when fishing is keeping nearly constant contact with the fly in motion. This means that it rarely touches the stream bottom on many rivers, and he believes the down-facing point tends to hook trout securely in their mouth. Advantage: hook facing down, right?

When I was fishing in Iceland during the summer of 2019, articulated streamers were on the menu for larger brown trout. At many locations the fish wanted slow retrieves, right off a shallow bottom of less than three feet. With lots to snag, I chose a pattern with an up-facing hook and was rewarded with some stunning fish. Think about the depth of water where you fish and the required streamer speed, then determine which way you

want your hook facing. Or you can be like me . . . tie them both ways and be prepared for anything!

Tailing materials play an integral role in the design of streamers, and the "easy out" is marabou. However, marabou takes on unneeded weight when wet, so for the rear hook, longer hackles are preferred. Schlappen is used in the EP Articulated Streamer; once wet, the schlappen feathers provide excellent movement while not absorbing a lot of water. This helps keep the back portion lightweight, allowing it to wiggle even in little current. Sticking with materials designed to keep the weight reduced, another favorite is EP Brushes, which also provide durability with their stainless-steel wire core. Since these do not absorb water, the patterns you tie are *much* easier to cast on the water. From a tying perspective, make sure you palmer the fibers as you wind forward, then use the dubbing brush when finished, releasing any fibers that may have gotten trapped. On to the connector!

There are many great connector materials, and I have tried lots with varying degrees of success.

One that allows for incredible movement between the two hooks is fly line backing, and I have a love/hate relationship with it. Some early struggles included determining the right distance to set the second hook so it wouldn't foul the front one, and once I felt I had achieved that, the fishing results were excellent . . . until that one day in Iceland.

Picture this: a crystal-clear pool with slow-moving water, set slightly downstream from a 25-foot waterfall. Nine feet at the bottom lies an Arctic char, but not your average size; this one is on the upper end of the spectrum, truly defining the term "fish of a lifetime." After extending my leader and tying on an EP Articulated Streamer, everything came together, from the cast landing in its proper lane to the fly sinking to the perfect corner of the pool. All that was left were a few strips of the line, and once made, the char charged and attacked the fly, just like my four-year-old son opening his Halloween candy!

After several strip sets, suddenly all pressure was gone, and once I examined the fly, it was apparent I hadn't tested my connection prior to the cast, as the rear hook was clearly in the jaw of the great char, visible to all of us. Let's return to fly line backing as a connector; where does the fault lie? Was the line dry-rotted, should I have checked the connection, did the char strike that aggressively? Maybe yes to all of them, but I have since switched to a wire connector, which has proven an upgrade in my opinion. Read on to learn an additional technique of securing the two hooks together.

✓ Tying Tip

Having two hooks attached to each other requires an understanding of various tying techniques. Be it cording thread, the use of adhesive, or even selecting varying connector materials, this junction is a critical one that could mean the difference between landing the fish of your lifetime . . . or left speechless on a pool in the middle of Iceland (I experienced both in a two-day period with articulated streamers). Aside from the techniques shared during this fly's procedures, another one to ensure a strong connection is to take the connector material and bring it through the ring eye of the

front hook. After doing so, lock the tag end of the wire to the hook shank through your typical procedures. Note that I was unable to do so with this pattern, based on the #6 hook size not having a major eye opening. As you begin to integrate larger hooks, you'll have that opportunity available, and threading the Beadalon (or other material) through the hook eye provides an additional layer of support to ensure that your two hooks stay together.

✓ Featured Technique

Cording thread is a technique discussed multiple times in this book, and for good reason: It's one that intermediate to advanced tiers use often with materials and threads. Spinning the bobbin provides tension to the thread, with that tension causing it to truly look like a bound rope . . . but don't spin too many times, as many threads will break. Finding that sweet spot takes experience; stick with it until you determine the perfect amount for your setup. As a tier, we're trained to nearly automatically create a smooth surface, though when you pair GSP Nano Silk with a plastic-coated wire, sliding enters the equation. Cording thread prior to lashing in Beadalon creates a rough surface that will grab on to the connector material; cording a second time to lash on top of the connector will allow the thread to bite in, helping to ensure your "fish of a lifetime" gets safely to the net!

✓ Materials to Consider

As materials are added onto the pattern, a primary concern is the profile. Articulated streamers represent baitfish, sculpins, crayfish, leeches, and even baby trout; each has a distinct body style and movement in the water. The EP Articulated Streamer is an overall generic pattern, and an easy way to modify styles is by varying the head; the version we tied has a sculpin helmet made by Flymen Fishing Company to represent sculpins, gobies, and baby catfish. In many of the waterways I fish, sculpins are in abundance, hence the profile choice. Other options include:

- Baitfish heads: Helping your pattern maintain a slender appearance, these heads provide weight and are simple to lock in place. Consider them an alternative to dumbbell eyes, which are another option when representing baitfish.

- Fish-Mask: Made of lightweight plastic, this product gives your streamer a baitfish appearance, yet allows it to ride higher in the water column. The lack of weight encourages body movement with the water currents, so I tend to use Fish-Mask patterns with a sinking line or tip. Finally, the clear plastic allows you to vary the head color based on the materials underneath.

- Natural materials: Using natural materials, such as deer belly hair and wool, gives you additional control of the finished look, though it requires more of a learning curve and may take increased time compared to using a ready-made head. Be aware of how different materials react with water, as deer hair tends to make the pattern ride higher in the water column; conversely, wool absorbs water and can be difficult to cast with smaller-weight rods.

✓ Fishing Suggestion

With many modern and classic books dedicated to streamer fishing, you have lots of material to sort through. My tips are straightforward but have contributed to personal success.

- First, understand that the rod tip helps provide movement of the pattern; simply stripping line in will not always provide the needed "trigger" for many predators. Focus on jigging motions, both upstream and downstream, then reconnect with the fly immediately.

- Slack is your worst enemy . . . along with a trout set! When a fish strikes, return the favor by making multiple strip sets. Even if you think the fish is gone, continue to strip-set (trust me on this one). When predatory fish are chasing, they tend to eat and swim with their prey, giving the angler the feeling that they rejected the pattern. Instead, their momentum keeps them moving in our direction, creating slack in the system. Strip-set quickly to retrieve that slack and set the hook, then enjoy the fight!

- Finally, two hooks versus a hook with shanks? If the latter, what about a trailing hook? These are two of the most common questions I receive about this and similar-styled patterns. It's my preference to use two hooks (where legal) due to the various nature of fish strikes. For example, I've noticed that more brown trout are hooked with the front hook, versus rainbows with the rear. This may relate to the types of water I'm fishing or differences inherent with trout, and the positives with two hooks definitely outweigh any downsides for me. With other species, especially saltwater, I see less need for multiple hooks, as most gamefish will devour the entire streamer.

Tying the Articulated Streamer

REAR HOOK

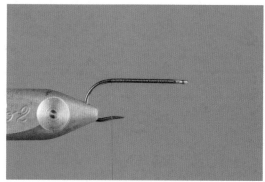

1. Start the thread with wraps behind the hook eye, then lay a base to the bend of the hook. Stop wrapping right before the shank begins to bend around.

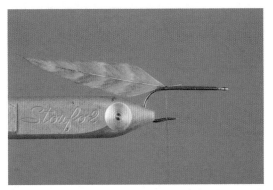

2. Select two schlappen (or similar) feathers and strip the butt fibers away. Lock one in place with thread wraps on top of the hook shank. Because of the quill shape, if you attempt to tie these in on the sides of the hook, they will turn as you wind. Instead, place this feather and the next on top to keep them secure.

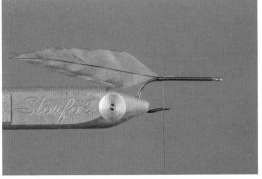

3. Lock the second schlappen feather in place, again on top of the hook shank. To guide this feather, I line them up by the tips, then use thread wraps to secure the stem. A consideration here is to have the feather lie either concave or convex, and that is a personal decision. Allowing them to splay apart, with the concave section (feather rear) facing out encourages more movement in the water. The reverse helps maintain a sleek baitfish appearance.

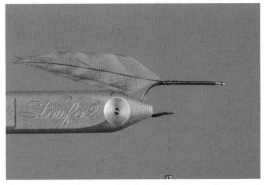

4. Cover the butt ends of the feather with thread wraps. Using a GSP thread in a fine diameter will allow you to cover the ends without adding unnecessary bulk.

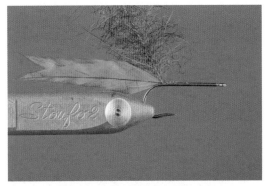

5. Lock in an end of the EP Brush with thread wraps, then wind forward to the eye of the hook. *Tying tip:* When locking in materials that are bound together, such as chenille and EP Brushes, I prefer to strip the material from the tag end, exposing the core. Securing the core will create less of a "bump" in that section of the fly and is an overall stronger connection.

6. Begin winding the EP fibers forward with nearly touching wraps. If you want more translucency in the pattern, allow spacing between wraps. As you wind, treat the material like hackle and palmer the straggly fibers toward the bend during each turn. Doing so will keep those forward-facing fibers from being trapped by the next turn of material.

7. Wind forward, ensuring that the EP fibers blend together with each turn. If any fibers are trapped under your wrap, simply unwind, pick them out, then continue toward the eye of the hook.

8. Stop winding just before the EP fibers touch the eye, then hold the tag end out over the eye. Secure the fibers in place with multiple thread wraps.

9. Trim the fibers with heavy-duty scissors. The EP fibers have a metal core that can damage the tips; cut farther down the inside angle of your blades. Build up the head with thread wraps, then whip finish and trim the tag end.

10. Apply Solarez Bone Dry to the brush, dabbing it inside the bottle to remove any excess. Touch the brush to the thread head and evenly apply the UV resin completely around the shank, covering all exposed thread wraps.

11. Cure the resin with a UV light for at least 10 seconds, or the time that the manufacturer recommends. When curing, be sure to shine the light on all areas of resin-covered thread, completely around the hook.

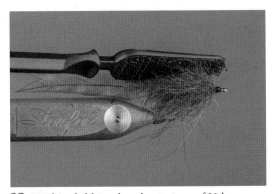

12. With a dubbing brush or piece of Velcro, begin to release any trapped fibers, moving the brush from the rear toward the eye of the hook. After you've brushed a section, the fly will appear very straggly.

13. Continue working to release the EP fibers with the dubbing brush, moving completely

around the shank of the hook. When doing this, be aggressive! The EP fibers are securely locked to a stainless-steel core and you're brushing to ensure that they are free to breathe.

14. The finished view of the rear hook. Note the overall simplicity of this style, but the key is that we've built lots of movement into the pattern. Remove the hook from the vise and set it aside.

FRONT HOOK

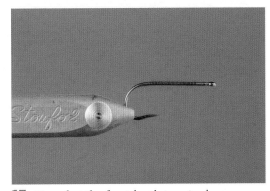

15. Note that the front hook is a size larger, which is intentional to maintain body consistency once the head is added. Place the second hook into the vise, then tie in the thread behind the eye. After laying five base wraps, spin the bobbin clockwise to cord the thread. Spiral wrap toward the bend, above where the barb would be. **Note:** By cording the thread, a rough surface has been created to help "grab" the wire during upcoming steps.

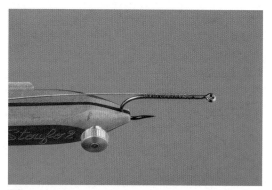

16. Cut a section of wire and lay it along the hook shank so its tip is about one quarter of the way back from the eye. Cord the thread again by spinning the bobbin in a clockwise direction, then secure the wire at the bend with heavy tension. Wire at the bend should be rising toward the top of the shank to ensure the trailing hook with the ring eye will ride with the hook point in the same direction as the front hook.

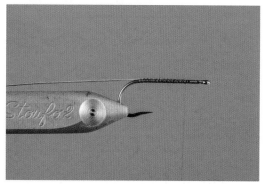

17. Wrap the thread forward to the eye of the hook, maintaining heavy tension. Wrap back toward the bend, ensuring the thread is corded and under pressure, then return the thread to the eye.

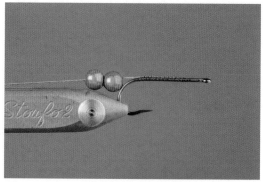

18. Slide two beads onto the single strand of wire, then push them up toward the hook bend. Don't overthink the colors; my favorites include red, black, orange, brown, and white . . . and red is used more than the other colors combined!

19. Place the tag end of the wire through the eye of the rear hook, then slide the hook behind the beads. A ring eye is used for this tutorial, and the wire is inserted through the top of the eye. As a result, both hook points will ride in the same direction.

20. Bring the tag end of the wire through the beads, then lay it on the hook shank. Cord the thread and advance rearward toward the bend. Cover the wire with corded thread while maintaining heavy pressure. Finish with the thread behind the hook eye and snip any excess wire. *Tying tip:* When determining the length for the wire connector, I prefer to have the beads touching the front shank. Next, I allow for a loop that is two hook eyes in length for the rear of the connector material. This spacing allows adequate room to adjust the beads and encourages movement of the rear hook.

22. Shine a UV light over the resin-covered thread for approximately 10 seconds, or based on manufacturer's recommendations. Ensure that you shine the light completely around the shank of the hook.

21. Seal the thread wraps with UV resin, being sure to completely cover with the brush. Remove excess glue by gently dabbing a piece of paper against the hook.

23. After curing, the wire covering the first shank should transition from a lower point near the eye to a gentle rise by the bend. Resin is also used to ease the transition from wire to hook by the eye. At this point the thread is resting between the wire and the hook eye.

24. Wind the thread rearward, stopping before the hook begins to bend. Measure a marabou feather so the fibers cover both beads and the rear hook eye. Using pinch wraps, secure the marabou feather in place on the far side of the hook.

25. Measure another marabou feather by aligning it with the one previously tied in, then secure it in place with thread wraps on the near side of the hook.

26. Wind the thread forward over the marabou butt fibers, leaving room near the eye for the head. Trim the butt ends of marabou, then return the thread to the rear of the hook.

27. Prepare the EP Brush by removing the fibers close to the tip of the core. Next, secure the EP Brush in place by the metal core with multiple thread wraps.

28. Wind the thread forward toward the eye of the hook, then begin to wrap the EP Brush around the shank of the hook.

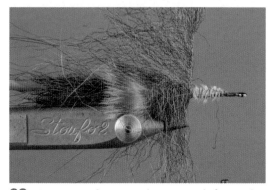

29. Continue advancing the EP Brush forward, palmering the material toward the rear of the hook with each wrap. Note in this picture how the metal core is visible as the fibers appear to sweep rearward; this technique prevents them from being trapped by the core.

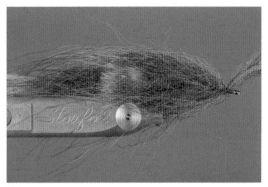

30. As the EP Brush is wrapped forward, be sure to use touching wraps with no visible gaps. The body will appear full as you near the hook eye.

32. Tie off the EP Brush with the thread by the hook eye, then trim the tag end close to the hook with scissors. If you've finished the brush close to the hook eye, consider folding it back over itself first before tying off.

31. Continue winding the brush forward, stopping a few hook eyes back. Reserve extra space by the hook eye to allow for use of a specific head. Vary your tie-off location based on what you plan on using for the head, which may include dumbbell eyes, a cone, or even simply thread. To ensure proper placement of the head, verify that you have space remaining by sliding the head in place. If the head slides back too far, consider building up the front of the section with additional wraps of EP Brush.

33. Brush out any trapped fibers by brushing aggressively toward the hook eye. Use the dubbing brush completely around the shank of the hook, working from the bend forward.

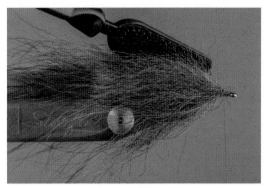

34. After ensuring fibers are no longer trapped, brush rearward toward the bend. The fibers should flow smoothly in that direction.

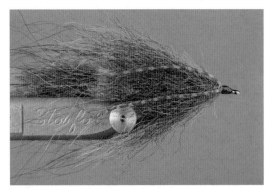

35. Tie in two rubber legs per side, then trim the legs so the tips line up with the rear hook eye. Build up the head with thread, then whip finish and cut the thread with scissors. The rubber legs can be positioned to ride along the side of the hook or slightly above. On this side, I have positioned them slightly above, which is preferred when the hook is riding point up. On the far side, the legs are tied along the side and are not as visible when viewed from the top.

36. Place a significant drop of superglue over the thread wraps and on the front materials. Be sure to coat everything, completely around the shank of the hook.

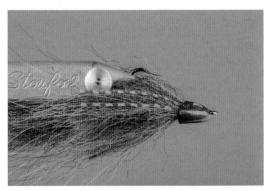

37. Place the head over the shank, pushing it snug against the materials. When using these sculpin heads, there is a weighted keel that will cause the fly to sit with the eyes up. In this case the fly will ride hook-point up based on the weight inside the head.

38. The view from the top. The head is pushed tight and leaves room in front for the thread wraps that will secure everything in place. The materials are all flowing rearward from the head and give a full-body appearance. Note how the rubber legs are pushed toward the sides of the pattern. Those on the top are not visible, as they are running along the side and would be more beneficial if this pattern was riding hook-point down. This demonstrates how a slight tying placement can alter the overall look of a pattern.

39. Secure the head by wrapping the thread directly behind the eye. Whip finish and trim the thread, making sure you've left an opening in the eye for your tippet.

40. Place UV resin over the thread wraps, then cure with a UV light. Be sure to place some resin in the gap at the front of the head to further secure everything in place.

41. Place drops of superglue in the eye sockets, then cover with the eyes. To further protect the eyes, cover with a dab of UV resin, then cure with the UV light. The eyes on this pattern are relatively lifelike, but I also encourage you to try some hot spot colors for them.

42. The finished pattern. When removing an articulated streamer with two hooks, be careful not to poke yourself! Note the full appearance of the finished streamer, which will get slender when wet. This pattern has a sculpin helmet, which matches local forage in the waterways I fish. Changing the head design makes this an easy pattern to vary based on your needs.

43. The overall length of articulated streamers can be intimidating to many anglers. Start with a size you deem manageable, then increase size according to your comfort level. The added length is an attractor for many fish, especially larger ones. Remember, the articulation encourages this fly to swim in a much more lifelike manner, aided by the schlappen tail. Finally, select a baitfish color scheme that you've previously had success with, then build the arsenal with the addition of more color combinations. Have fun with one of my favorite streamer patterns!

Extreme String Baitfish

The Extreme String Baitfish pattern offers just the right amount of translucency and features some newer materials in its design. Having a mix of imitative and suggestive forage-fish imitations in your box gives you lots of flexibility and choice when on the water.

With so many options for streamer patterns, sometimes using Craft Fur feels like taking the easy way out. But the material looks incredible on baitfish imitations and is simple to use. Lots of color options exist, making this an excellent starting point for this fly style. On the Extreme String Baitfish, you'll learn the basic tying techniques for Craft Fur, and I'll share other modern materials that absolutely deserve a place on your flies today.

Before jumping into tying this style of pattern, know that there are different versions of Craft Fur on the market, and you can find the material online, in fly shops, and yes, in craft stores. Like most things in fly tying, not all products are the same, with varying diameters, lengths, and even number of fibers per square inch. Add to this an

Extreme String Baitfish

- **HOOK:** #6 Hanak 970 BL
- **THREAD:** White Semperfli Nano Silk 18/0
- **WING AND TAIL:** White Extra Select Craft Fur
- **FLASH:** Pearl Midge Flash
- **BODY:** Cream Semperfli Extreme String
- **HEAD:** Fish-Mask #3
- **EYES:** Fire Fish-Skull Living Eyes 3.0 mm
- **RESIN:** Solarez Bone Dry

insane number of color options, and you can get just a tad overwhelmed. I recommend starting with a few base colors, especially those that imitate baitfish in the waters you fish, then expand as needed. In my Craft Fur "collection," favorite colors include white, black, and olive, and there are even grizzly styles on the market today that have found their way onto my bench.

Returning to the conversation with Craft Fur, consider treating the material like deer or elk hair. This involves trimming with serrated scissors, combing out the butt ends due to excess fibers, and even straightening the tips in your hand. Stacking is unnecessary, as the varying lengths help create a natural flow toward the rear of the hook. Next, before locking the fur in place, I suggest a thread base, which will prevent them from sliding around the shank of the hook. Finally, notice that I placed the tips forward, which eventually were folded back. This technique helps create minimal buildup at the eye of the hook, instead forming a rounded head on the fly, like its baitfish doppelgänger. Another advantage is that the fibers will splay out from the shank when folded back, which helps guarantee movement when they're wet, and fight water resistance as you are stripping the streamer toward you.

Tailing options are discussed more below, but in general, my preference is to use Craft Fur, as it blends nicely when combined with its twin from the head. That is an important piece for me when imitating baitfish, that the flow of the overall profile transitions smoothly from head to tail, which can be interrupted with the inclusion of differing materials. Notice that a few pieces of flash were added, a deliberate choice since the body material has fine sparkle, too. Subtle flash is found on most of my baitfish flies. The Extreme String meets the need of an easy-to-use material with a slight amount of "twinkle" integrated.

As new materials come onto the market, tiers benefit the most. Extreme String, offered by Semperfli, is both easy to use and comes in a variety of colors, perfect for tying flies to catch predatory fish in both freshwater and saltwater situations. If you're like me, your tying area undoubtedly houses every color created of your favorite products, encouraging creativity at the vise.

Continuing with the body fibers, Extreme String fibers are attached to a core, similar to chenille, which helps make them more resilient. Streamers can get destroyed by aggressive fish in a short amount of time, so I tend to choose body materials that can withstand punishment, especially from teeth. Another option is a simple thread body (covered with UV resin or superglue), as the Craft Fur head will allow for some translucency, mimicking many minnows found in waterways around the world. This also helps ensure an overall slender profile. Finally, if you're searching for a stronger material while still wanting to achieve bulk and flash, consider dubbing brushes created with steel wire. For an example of these in action, check out the EP Articulated Streamer in this book.

When discussing baitfish imitations, eyes are an essential part, as it's my belief that they can often trigger a strike. For this fly, I introduced a product used to secure the eyes and push back the head, known as a Fish-Mask. This approach is simple and straightforward, while providing an authentic and overall slender look to your pattern. Experiment with the head style you prefer, from dumbbell eyes to even Craft Fur folded back with eyes glued directly to the fur. Speaking of eyes, there are lots of options for color and I love them all . . . as long as they're red. Sealing everything with UV resin makes this a durable choice and one of my go-to streamer styles for fishing today.

✓ Tying Tip

Using the Fish-Mask at the head makes securing the eyes a cinch, yet remember that it has the potential to come off the pattern, especially after multiple uses with aggressive fish. Here are some thoughts to help things out:

- Before finishing the head, I prefer the body materials to be significantly built up, as we want the Fish-Mask to be snug. Crowding the hook eye is typically a negative, but keep the materials close to that edge, as they will help bind everything together once adhesive is applied.
- Dab resin all over the materials prior to sliding the head on top. Once it's securely in place, curing the resin is simple, as the UV light

penetrates the semitransparent mask. If you're able to twist the head after curing, then you know either more resin was needed or the materials underneath weren't secure to the hook shank.

- Superglue is another excellent option, though it takes a little more time to set in place. The same thoughts apply for traditional resin.
- Thread in front of the head helps to finish the fly, then the use of UV resin to cover the eyes and those forward thread wraps really keeps things protected.

✓ Featured Technique

When I first saw Craft Fur used on baitfish imitations, I was impressed with its ability to provide sleekness to the imitations. Prior to any experience tying with it, I kept wondering why so many tiers folded the material back versus tying it like a traditional wing (butt ends facing the hook eye). An early attempt was to integrate this into a Black Ghost variation, replacing the hackle or marabou wing with Craft Fur . . . and then I realized that two main issues existed:

1. Craft Fur butt ends built up quickly, making the front appear bulky.
2. The head fibers typically stayed tight to the shank, not flowing gracefully with the water when wet.

From that point forward, I chose to place the material with the tips tied over the hook eye, then folded back toward the tail. This helps solve the above issues, while also giving the pattern a more rounded shape to mimic baitfish. After folding back, be sure to create a substantial thread dam in front of the fur, which will ensure that the material remains pointed rearward.

✓ Materials to Consider

There are many options for tails on your baitfish imitations, with this one offering a Craft Fur and flash combo. Folding a few pieces of Midge Flash help continue the slight amount found within the body, and that's something I prefer in most of my streamer patterns. In general, the tails on this style of fly should offer a continuation of the body . . . plus lots of movement! With many choices available to you as a tier, here are some considerations:

- Feathers. Adding feathers encourages movement with minimal bulk on baitfish imitations. They also absorb little water, making them easier to cast.
- Flash. How much flash added is completely up to you, and I've seen some tiers go CrAzY with a tail made completely of flash. Doing this is like incorporating a "hot spot" on the fly, and I recommend trying some hi-vis colors that will guarantee the fish's attention, especially in off-colored water.
- Craft Fur. Combined with the fur folded back from the head makes this baitfish imitation blend together nicely, as done in the example I shared here. Another option is to vary the color, which gives an opportunity to create an obvious darker top and lighter bottom, more accurately mimicking various baitfish.
- Rubber legs. Nothing says movement and vibration like a couple sets of rubber legs, and a common tying technique is to fold them around the thread and add them to other tailing fibers. Occasionally, I'll also add a set closer to the head, and a tip is to point the rubber legs forward toward the hook eye. Once in the water, the rubber legs get pushed toward the tail when stripping the fly in; however, they are encouraged to "fight" the water and vibrate each time that motion is paused. Continue to think about ways that materials interact with the water and incorporate those ideas and thoughts into your patterns.

✓ Fishing Suggestion

Streamer fishing is a very productive method, especially when you commit to the process for a full day . . . or entire fishing season! With so many choices of fly types available, it can be overwhelming to know which is most effective. Follow these general tips for a starting spot, then build based on your experiences fishing.

- When on the hunt for big fish and covering lots of water, I tend to select articulated streamers,

as they offer an incredible amount of movement. Fish them in the upper water column and attempt to generate lots of movement with a softer rod tip. Flies with nonabsorbent materials are preferable when fishing all day, as they are much easier to cast.
- Once the search is narrowed to specific locations, especially pocket water, smaller streamers are my first choice, such as the Extreme String Baitfish and Mini Jig Bugger. They offer a great profile, accurately mimic local prey in the area, and can be fished at varying depths depending on their weight.
- The first color I fish for the day can greatly vary, based on water conditions, time of day, previous experiences . . . or even a gut feeling. In addition to my base colors of white, black, and olive, other favorite ones include lighter colors like yellow, orange, and chartreuse. When contrast is built into a pattern, fish tend to respond positively, and color combinations such as pink and black work well, especially in off-colored water.

Tying the Extreme String Baitfish

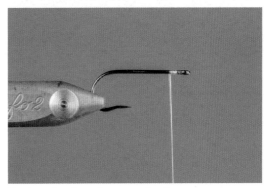

1. Start the thread one eye back, wrapping rearward, then trim the tag end of the thread tight against the hook shank.

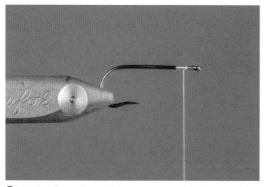

2. Make four thread wraps rearward, then cord the thread by spinning the bobbin clockwise. Once corded, spiral wrap the thread forward. Note how the corded thread appears thinner but denser, sitting higher than the initial wraps.

3. Trim a clump of Craft Fur and hold it by the tips. Note how the butt ends are full of excess material, which makes the tie-in process difficult by adding unnecessary bulk.

4. Holding on to the Craft Fur tips, use a dubbing brush to comb out the excess material. Move the brush from the tips toward the butt

ends and make multiple passes. Save that excess material for other creations!

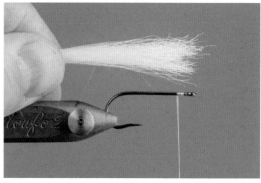

5. Once the excess material is removed, the Craft Fur will be uniform from tips to butt. Measure the craft fur to be approximately twice the length of the hook shank.

6. Once measured, turn the material so that the tips are facing forward over the eye of the hook. Cord the thread by spinning the bobbin in a clockwise direction, then lock the Craft Fur in place at the eye with pinch wraps. The tips should be approximately twice the length of the shank, and when tying in, ensure that the Craft Fur is on the top of the hook shank.

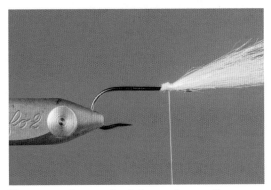

7. Trim the butt ends of the Craft Fur, allowing a slight ramp toward the rear of the hook. If you have a rotary vise, inspect the hook from various angles to ensure that all the material is flush on the top of the shank.

8. Invert the hook in the vise . . . and be careful of the hook point! The vise in the picture is a Stonfo Transformer with the streamer head, which is a straight set of jaws that allow me to rotate without removing the hook.

9. Repeat the tying procedures with a second clump of Craft Fur. Trim it to length first, then

remove excess material by brushing it out. Measure to the same length as the first tied-in clump by aligning the tips, then lock in place after cording the tying thread. This clump of Craft Fur should be tied on the bottom of the hook shank.

10. Trim the butt ends of this clump. Note that I left the material longer before snipping. Instead of having a large bump of materials near the front of the pattern, by trimming the second clump longer you'll create a smoother transition. Cover the butt ends with thread.

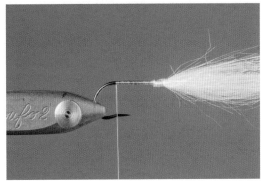

11. Invert the hook in the vise, then continue wrapping the thread back to the approximate location of the hook barb. For barbless hooks, I prefer to wrap one-third of the way back from the hook point. Prepare another piece of Craft Fur for the tail, cutting a section approximately the same length as the hook shank. Remove the excess from the butt ends and line it up over the hook shank.

12. Lock the tailing fibers in place with thread wraps, then wind forward to trap the butt ends of the Craft Fur. Trim away any excess, finishing the material so the butt ends of the Craft Fur all come together on the hook shank. Return the thread to the rear of the hook, covering the Craft Fur with a layer of thread.

13. Fold four pieces of flash over the thread, holding them by their tips with your left thumb and pointer finger. As you advance the thread around the hook shank, hold the flash taut, with the ends over the hook shank. When adding flash, I prefer to go with a more subtle appearance. Use this as a starting place, then adjust based on personal preferences and experiences on the water with other "flashy" patterns.

14. Continue wrapping the thread around the hook shank but hold the pieces of flash above the hook, preventing them from spinning around. The tension of the thread will pull down on the flash, securing them to the top of the hook shank. Make another couple wraps to secure the flash in place. They will now be seated on top of the tailing fibers.

15. Cut a long section of Extreme String, then place one end on the hook shank. The other end of the string should be resting over the tail. With a pinch wrap, lock in the Extreme String at the base of the tail, then wind down its butt end, covering it with thread. Once secure, advance the thread forward toward the eye of the hook, stopping at the base of the Craft Fur.

16. Begin winding the Extreme String forward with touching wraps, ensuring the flash and fibers face up, similar to hackle. While winding, slightly palmer the materials toward the tail of the hook, which helps prevent future wraps from trapping them underneath.

17. Continue winding the Extreme String forward, keeping tension on the material as you wrap. Note in this picture how I have continued to preen the fibers both up and back, keeping a clear path forward.

18. Conclude winding the Extreme String once you've reached the base of the front wing. Finish

with the last turn about to touch the wing. If you overwrap and the Extreme String is wrapped over the wing, it will make things very difficult to finish in later steps. Secure the Extreme String with thread wraps.

19. Once the Extreme String is locked in place, snip the tag end with scissors.

20. Pull back both clumps of Craft Fur fibers, then carefully wind the thread between the two sections. Once reaching the head, form a thread dam, gently building a slight ramp from the eye to the wing. *Tying tip:* To help with this process, gently moisten the Craft Fur fibers, which will keep them in place going rearward. This is a technique I use with many synthetic and natural materials that are otherwise hard to work with.

21. Whip finish the thread at the head, then trim it tight against the fly. While still holding your scissors, trim the flash tail so it extends just past the tips of the Craft Fur.

22. Prepare to apply UV resin onto both the thread head and into the edge of the Craft Fur. I prefer to hold the nozzle by the hook eye and push the resin rearward; doing so in reverse may cause excess resin to cover the hook eye. The head being used on this pattern is clear, which allows the UV resin to be cured through itself. If you plan on using an opaque head, superglue is the recommended adhesive.

23. Apply the UV resin with gentle pressure, building up a portion around the hook shank

and over the front section of the Craft Fur, as that material will also come into contact with the head.

24. Slide the head into place, pushing it back far enough to ensure there is room to make thread wraps between the head and the eye of the hook. If you did not apply enough resin to grab on to most of the head, quickly remove it and add additional resin. Once the head is in place, cure the resin underneath with a UV light.

25. Build up thread in front of the head, applying enough to form a taper toward the hook eye. These thread wraps will help keep the head in place against the fly. Whip finish, then trim the thread tight against the fly.

26. Place the eyes on both sides of the head. Those used for this pattern have a slight adhesive on the rear side, which holds them enough to get them into position.

27. Using the rotary function of your vise, turn the fly 90 degrees for access to the outside of one eye. Apply resin, using enough to completely cover the eye. Cure the resin on that side of the head with the UV light, then rotate the vise to repeat this same procedure with the other eye.

28. Apply UV resin to the thread wraps in front of the eye, creating a smooth taper that flows from the head to the hook eye. Cure the resin with the UV light. *Tying tip:* Be sure to keep resin out of the hook eye, and if any gets in, run your bodkin through the eye to remove the excess. For smaller hooks, try running a fine hackle stem through to clear the eye.

29. With the Extreme String underneath, this fly allows translucency through the body, similar to many baitfish in our waterways. Aside from using basic colors of Extreme String and Craft Fur, I also like to use permanent markers and add additional features, such as red gills, a lateral line, and a darker back. These small details take little time but can be the trigger to help make this fly even more effective. In the end, this all-white version is a favorite of mine, and one I have used to catch a variety of species during all four seasons of the year.

Mini Jig Bugger

The Mini Jig Bugger is an elegant and simple tie, with techniques that can be applied to many of your favorite patterns. In this tying tutorial, we will discuss a way to make your standard nymph hooks act immediately like a jig, all with the "incorrect" placement of a slotted tungsten bead.

What fly-tying book would be complete without some form of a Woolly Bugger, right? The original streamer is cemented as a fly for the ages, commonly used in tying classes as "the fly" beginners are initially taught. This is for good reason, as many of the materials and styles within that pattern can be translated to others . . . plus it's a proven fish-catcher! Within the Mini Jig Bugger, intermediate techniques are demonstrated, plus you'll gain tips to benefit other staple patterns at the bench.

This pattern shares more ideas than initially meet the eye, including an inverted slotted bead and a simplified body created from a brushed-out dubbing loop. My box is loaded with patterns incorporating these two techniques, and yours

Mini Jig Bugger

- **HOOK:** #12 Hanak 230 BL
- **BEAD:** Silver 3.5 mm slotted tungsten
- **ADHESIVE:** Fly Tyers Z-ment
- **THREAD:** Black Semperfli Nano Silk 18/0
- **DUBBING:** Black/red Arizona Simi Seal dubbing

should be, too. The Mini Jig Bugger is a perfect learning pattern, and with slight modifications can replicate baitfish, leeches, and even stonefly nymphs. *Fly Tying for Everyone* encourages you to grow with your tying, and if the traditional Woolly Bugger is still a close friend, this chapter should help bring the pattern to a new level.

Good things come in small packages, and this mini streamer starts with a traditional down-eye nymph hook. By inverting the slotted bead, we're encouraging the fly to ride hook-point up, like a jig hook. Options abound for those desiring them! Want to guarantee the pattern plunges to the depths quickly? Add a second tungsten bead behind the first, something that I do on many nymph and streamer patterns, especially those where added bulk is welcome closer to the eye. If a slender profile is the goal, I'll substitute wire behind the front bead, typically in the range of .010- and .020-inch diameters. Remember: Weight near the eye encourages a jigging motion, especially when stripping this pattern with slow twitches.

Throughout this book, you've been exposed to many tailing options, including marabou, Craft Fur, rubber legs, Krystal Flash, feathers, and Coq de Leon . . . and they all work great with this style of fly! However, professional fly tier and friend Justin Aldrich ties a similar version, and one of the tricks he encourages is to tie the pattern sans tail, then brush out the dubbing to extend past the bend . . . voilà! This is a fun technique, and I encourage you to experiment with various dubbings, as those with lengthier fibers brush out for a longer tail. Before we get to that point, let's talk about what happens with the body.

Creating a dubbing loop is a critical step for nymph patterns that demand a buggy look, especially around the thorax to create the appearance of legs. With baitfish imitations, the lengthy fibers are a welcome addition, encouraging movement when in the water. From the tying perspective, a dubbing loop is a secure method of adding depth to flies, giving them bulk near the outer tips that comes with little weight. In short, dubbing loops are important for a variety of tying, as illustrated throughout this text.

After spinning the dubbing loop, we must take special care winding the dubbing rope forward. Knowing that we're selecting dubbing with straggly fibers, it's easy to trap them underneath the wraps, quickly deteriorating the translucency and movement we're building into the fly. With each wind, ensure that the fibers are carefully pulled away from the dubbing rope toward the rear, like palmering hackle. This technique prevents the forward wraps from grabbing fibers and maintains a naturally smooth flow toward the fly's tail.

A quick brush of superglue on the thread helps secure things during the whip finish, then it's on to an essential step, brushing the fly. Be sure to free any trapped fibers with a dubbing brush or something abrasive, such as the hook side of Velcro. If fibers are longer than 2 inches, I like to brush them after spinning the dubbing loop, then again once everything is sealed following the whip finish. For the former, ensure your loop is spun tight, or else fibers will be pulled out, causing the body's profile to appear uneven. A less abrasive method is to run a fingernail along the dubbing rope after spinning, which will free many straggly fibers before a second spin. Once everything is locked into place, it is all right to be slightly aggressive when brushing

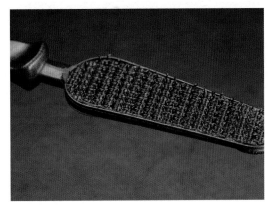

A dubbing brush is a tool that can be found both on my tying bench and in my travel bag, waiting to loosen trapped fibers, add depth to a pattern, or even create straggly abdomens or thoraxes from tightly wrapped dubbing. When using a dubbing loop, this tool is at the ready as I near the fly's completion.

both toward the eye and the bend, as those longer dubbing strands really help take the place of the Woolly Bugger's hackle fibers.

Take a step back and examine this no-nonsense pattern tied with minimal materials. Each step is very intentional and part of the pattern's design, heeding the mantra of "less is more." It doesn't always seem that way in tying or fly fishing, both of which have been known to overcomplicate things, to the chagrin of others. Don't fall for that trap . . . unless you enjoy the challenge; instead, examine your favorite and most effective patterns and begin to question the materials and techniques within each. Are they being tied to simply maintain a status quo, or can they be improved upon? My challenge to you is simple: Experiment at the tying bench and find ways to ensure your flies will perform at their highest possible level.

✔ Tying Tip

There are many ways to lash dubbing onto a hook, including a dubbing noodle, touch dubbing, split thread, and a dubbing loop. When going for a buggier look, I tend to select the latter, and this technique is a time-saver to both build a body and help spin hard-to-dub fibers effectively. Let's take a quick look at some ways these methods differ:

- Traditional dubbing noodle. Go with this method for natural fibers and easy-to-dub synthetics when desiring a tighter dubbed body, especially for many dry-fly and nymph patterns.
- Touch dubbing. Want a bit more of a straggly body? Here's a technique to try, which involves first applying wax to the thread, gently touching dubbing to the wax, then spinning the bobbin to trap the fibers. Once wrapped in place, this technique gives the appearance that you lightly brushed the fibers out, and this is a preferred method for the thorax of various caddisfly pupae imitations.
- Split thread. Similar to above, this is a fast way to create a dubbing loop and is also a favorite for other materials, such as CDC. Steps include uncording the thread, flattening then splitting with a needle or bodkin, followed by the material placement inside the loop. Once the bobbin

gets spun, you're off and running . . . but (there's always a but!) not all threads are created equal. Those that consist of many strands are much easier to split. If you're having problems, be sure to try a couple different types of threads before giving up.
- Dubbing loop. My favorite for creating the buggiest look with dubbing. This technique can be used with a variety of fiber lengths and materials, including CDC and even deer hair. Ready to take your tying to the next level? Try mixing multiple types of dubbing or materials inside the loop!

✔ Featured Technique

Be careful when reading the title of this chapter, as we're not using a jig hook for this fly. Instead, we are starting with a typical nymph hook and inverting a slotted tungsten bead. By doing that, the fly is encouraged to ride hook-point up when in the water, and there is a significant gap that allows for a better hookset. This technique, originally shared with me by Fly Fishing Team USA member Josh Miller, pays dividends, as I now only purchase slotted tungsten beads for most hooks I use, not just jig hooks.

✔ Materials to Consider

When it comes to dubbing, the choices are staggering . . . and growing exponentially! All right, maybe it's not that crazy, but there are many types to choose from, with some having a major impact on your patterns. Let's briefly discuss some favorite types that are preferable for many of today's fly-tying styles, plus an easy-to-use alternative:

- Ice dubbings. These synthetic dubbings are best when you're looking for something that will provide flash to reflect a lot of light, and there are many color choices. This category can be wide-ranging, all the way from metallic types to Antron blends. Many ice dubbings feature longer fibers (an advantage of dubbing loops), though a downside is that these can be difficult to use when forming dubbing noodles.

- Natural fibers. This category can have many subsets, including rabbit, squirrel, beaver, muskrat, and more! Many natural fibers dub onto thread easily, and I especially like those that include guard hairs for a buggier look.
- Superfine. This dubbing is commonly labeled for use with dry flies and consists of short synthetic pieces. Some brands will apply waterproofing to the dubbing, helping with additional flotation. Overall, this is an easy-to-use type of dubbing, especially when forming tight and slender bodies.
- Blend. With many of my natural nymph dubbings, I like a little added flash and will blend a small amount of ice dubbing into them. A popular blend that you can purchase includes Hare's Ear dubbing with small pieces of Antron mixed in, creating a buggy look with just enough sparkle to grab a fish's attention.
- Weighted. I had to throw an oddball in here, as I occasionally will tie steelhead flies with dubbing that is made of metal fibers. It dubs surprisingly well and helps add bulk in places without significantly increasing the mass.
- Yarn. One of the easiest methods to apply the look of a dubbed body is to tie in a strand of yarn that simply gets wound forward, a technique popularized by Frank Sawyer's Killer Bug. New products in my arsenal include two by Semperfli: Dirty Bug Yarn and Floating Poly Yarn. Though the tying process is similar for both, the Dirty Bug Yarn is intended for nymphs and emergers, as it will absorb water, while the Floating Poly Yarn is ideal for dry flies and comes in a variety of colors to match floating insects around the world. If you're feeling creative, furl multiple strands of different-colored yarn together for a unique appearance. Another fun variation is to tie in yarn like a dubbing loop, then select straggly fibers to place inside prior to spinning. Great choices to insert should move well in the water and even provide flash, such as CDC or ice dubbing.

✓ Fishing Suggestion

The Woolly Bugger has been such a popular pattern because it's easy to tie . . . and catches lots of fish. The Mini Jig Bugger is a close relative for those same reasons, and it's tough to fish this style of fly wrong! Try these methods first:

- Dead drift. Cast this style of pattern upstream and let it drift with the current drag-free. Encourage the fly to ride within one seam, which helps the Mini Jig Bugger appear lifelike.
- Jigging. As the fly is being swung, retrieved, or drifting freely, jig the rod tip with slight twitches . . . then hold on! This is an effective technique to employ both in moving and still-water situations.
- Quick strips. Since weight is consolidated at the eye, retrieving with occasional pauses will incur up-down movements, encouraging the fly to appear as if it's potentially wounded for that large trout to attack. Speeding up the line will help to mimic the Bugger darting away from its predator.
- Seesawing. Cast down and across, encouraging the fly to swing out below you. During that swing, seesaw the line by giving a 12-inch strip, then allow that amount to return to the drift. Continue to retrieve and release, as this technique helps the fly to stay at a desired depth during the drift.

Tying the Mini Jig Bugger

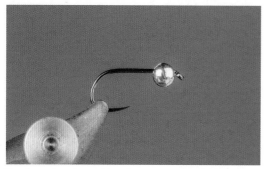

1. Place the slotted bead on a standard down-eye nymph hook. Note how the typical placement on this hook encourages the bead to decrease the hook gap, which we will fix in the next few steps.

2. Invert the jaws of your vise with the rotary feature. Now note that the slotted bead is also reversed and that less tungsten is showing in the hook gap.

3. Slide the bead to the rear of the hook, allowing working space closer to the eye.

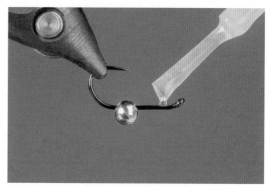

4. Place a drop of superglue behind the eye of the hook. If you've applied too much, dab away the excess with a paper towel or the edge of a piece of paper.

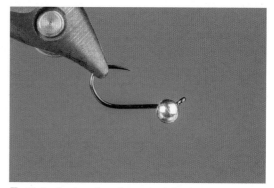

5. Slide the bead to the eye of the hook, ensuring that it remains inverted with its main mass below the hook shank. Allow 30 seconds for the superglue to dry, or you can also jam wire behind the bead to help it maintain its position.

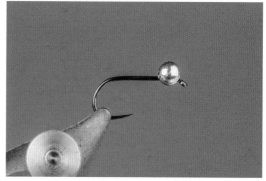

6. Rotate the vise back to its original position and the bead should stay inverted. Now note the significant gap located between the bead and the hook point. By inverting the slotted bead, this standard hook will tend to ride point up in the water, snagging less. More importantly, the wider gap will help you hook a greater number of fish. *Tying tip:* When you want additional weight on this pattern (and others), add a second tungsten bead behind the first. If you do this, be sure to determine a system to identify the heavier patterns or place them in a designated section of your fly box.

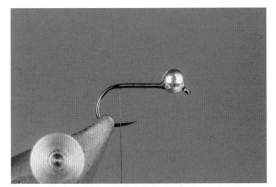

7. Tie in the thread behind the bead, creating a dam heading toward the shank. Trim the tag end of the thread, then wrap it to the rear, stopping at the hook of the shank.

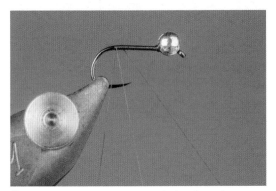

8. Create a dubbing loop with the thread, locking it in place by wrapping both behind and in front of the loop. The dubbing loop should be approximately 3 inches in length, depending on the size of your hook shank and how much material is desired.

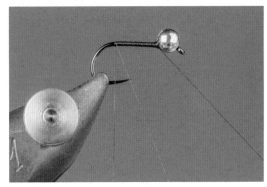

9. Advance the thread to the bead; if your vise has a bobbin cradle, allow the thread to rest there. Note in this picture that the thread is angled slightly forward, which helps keep a clear path for the dubbing loop when it is brought forward.

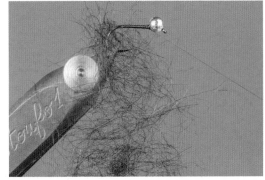

10. Place the dubbing clump into the loop, allowing the fibers to extend about 3/4 inch off each side. Ensure that the dubbing is evenly spaced, and slide the top section of dubbing up so it is tight to the shank of the hook. Selecting the perfect amount of dubbing is something that is learned over time, and for the #12 hook in this example, three small clumps was ideal.

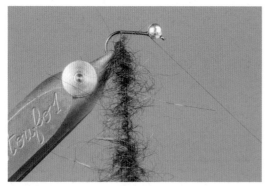

11. Spin the dubbing loop until you notice the fibers begin to lock each other in place. In this picture, you can see the red and black fibers are intertwined, but with quite a few trapped underneath.

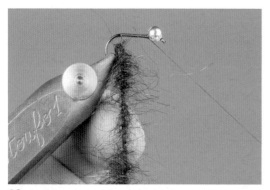

12. Holding on to the dubbing loop, use your fingernail to gently brush out the fibers in the dubbing loop. A dubbing brush or Velcro may be used, but many fibers can be pulled away from the loop.

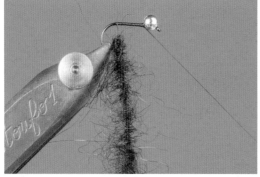

13. At this point the fibers are now released and ready to be wrapped forward. Before doing so,

look closely at where the dubbing loop meets the hook. In this picture, there is a slight gap where the dubbing got pushed down while using my fingernail. Simply slide the dubbing back up if this has occurred.

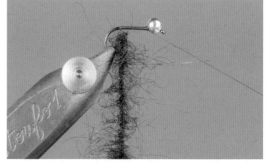

14. The dubbing has been pushed up the dubbing loop and is now resting closer to the hook shank.

15. Begin to wind the dubbing loop forward with touching wraps. Before each turn, fold the fibers rearward, as shown here. Hold them in this position as you complete the wrap, ensuring that the fibers don't get trapped underneath a new turn.

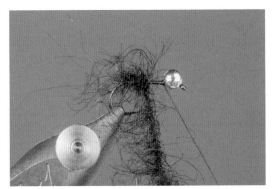

16. Continue winding forward, and the stragglier you get this to look, the better! Keep your wraps close to each other to create more bulk; allow some spacing for a more translucent appearance.

17. With each turn, continue to pull the fibers rearward from the dubbing loop. To help with this process, gently moisten your fingers, which in turn helps the dubbing fibers remain splayed outward. Wrap forward until reaching the bead.

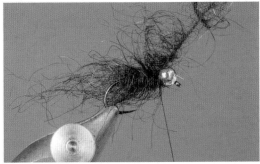

18. Once reaching the eye with the dubbing loop, secure in place with three wraps of thread. At this point, I am holding the dubbing loop with a pair of hackle pliers, which allow for greater control when winding.

19. Whip finish the thread directly behind the bead. Do your best to avoid trapping any straggly fibers under the thread.

20. Trim the thread close to the bead. At this point, if I want to add a quick hot spot of thread in a hi-vis color, I will build up several wraps directly behind the bead, then brush superglue onto the thread prior to a whip finish. This pattern used a darker thread, but you can always go with a fluorescent color, which would be hidden beneath the dubbing until the last few thread wraps at the head.

21. With a dubbing brush, brush aggressively from all sides to remove any trapped dubbing fibers. My initial movements are toward the bead of the fly, giving it an almost Tenkara-style look.

22. Continue brushing forward around the body of the fly, especially underneath, between the body and hook point. Notice how straggly the fibers have become with just a few brushstrokes.

23. Reverse direction and begin brushing toward the rear of the hook. This dubbing consists of longer fibers that can be brushed out to create the appearance of a tail. *Tying tip:* By experimenting with different dubbings, you can create varying looks, such as "spiky" due to guard hairs, or even incredibly flashy, caused by strands of tinsel integrated into the dubbing. Though the Mini Jig Bugger encompasses a body formed by a dubbing loop, think about how you can integrate this technique into some of your favorite emerger, nymph, and streamer patterns.

24. Lots of room remains for creativity, including with the tail. From marabou to a flashy material, color choices range from natural selections all the way to hi-vis options. Thinking about other body options, while dubbing is a versatile material, be sure to try different materials in the loop, such as CDC, flash, yarn, or even deer hair. Most importantly, incorporate this style of streamer into your tying and fly fishing, as it's a competent replacement for the trusted Woolly Bugger.

TECHNIQUES INDEX